An Introduction to Asset Corrosion Management in the Oil and Gas Industry

Third Edition

Ali Morshed

NACE INTERNATIONAL

NACE International
The Worldwide Corrosion Authority

Printed in the United States of America

ISBN: 978-1-57590-407-8

NACE International
The Worldwide Corrosion Authority
15835 Park Ten Place
Houston, TX 77084
nace.org

Dedication

This book is dedicated to Mr. Farhad Agah (the second son of Mrs. Manijeh Shahriari and Mr. Mehdi Agah) for his impeccable contributions to the Iranian cement and pistachio industries in the Kerman Province.

اندیشه نیک
GOOD THOUGHTS
کردار نیک
GOOD DEEDS
گفتار نیک
GOOD WORDS

Contents

Introduction to the Book and Its Author

The first edition of this book was published in May 2012, and in less than two years it had become a NACE International bestseller. Even though it was intended mainly for the upstream oil and gas sector, it was also very well received by the midstream and downstream sectors. Furthermore, it was also taken up by various non-hydrocarbon industries as a guide for personnel training and a framework for implementing corrosion management strategies in the field.

This warm reception within the hydrocarbon and the non-hydrocarbon industries, as well as within academia, encouraged the author to publish the second edition in October 2016. Similar to its predecessor, the second edition was also very well received by various hydrocarbon and non-hydrocarbon industries alike, and has become their main corrosion management reference book.

I have personally known Ali Morshed for more than 10 years now. Ali started his career as a corrosion engineer in the U.K.'s offshore oil and gas industry and later began working internationally. Ali possesses a great passion for writing and teaching and has eagerly demonstrated this passion through publishing five corrosion management books (including this one) through NACE in the past eight years, as well as providing numerous training courses. Considering his various endeavors in the area of asset corrosion management (i.e., provision of training, consultancy, and writing on the subject) there are very few individuals who can claim to match his accomplishments.

Congratulations with this new edition of the book!

—Dr. Torben Lund Skovhus
VIA University College, Horsens, Denmark

About the Author

Ali Morshed received his M.Sc. degree from Imperial College London in 1997, and his Ph.D. from University College London in 2002–both in corrosion engineering. He then moved to Aberdeen, Scotland and worked for a number of different U.K. companies from 2002 to 2013 as a corrosion engineer. In 2013, he joined Saudi Aramco for two years before becoming a consulting corrosion engineer in the latter half of 2015.

He has so far worked on asset integrity management, corrosion engineering, and corrosion management projects both with engineering and operating companies in the United Kingdom's North Sea, North Africa, the Persian Gulf Region, and South Asia.

Dr. Morshed is the author of the NACE International book, *An Introduction to Asset Corrosion Management in the Oil and Gas Industry* (2012), whose 2nd edition was published in 2016, and the 3rd edition (this current book) in 2020. He has also authored two more books on this subject for both the hydrocarbon and the non-hydrocarbon industries. The first one was published in July 2017 and the latter in May 2018.

In addition, he has been providing corrosion management training for the hydrocarbon industry since 2005, and for the non-hydrocarbon industry since 2016. Dr. Morshed has also taught the same topic to postgraduate students in Robert Gordon University in the United Kingdom.

He is a regular contributor to the NACE International's *Materials Performance* (*MP*) magazine on the subjects of corrosion management and microbially induced corrosion (MIC). Currently, he spends the majority of his time providing corrosion management services (i.e., training, auditing, and consultancy) for both the hydrocarbon and the non-hydrocarbon industries.

Author's Notes on the 3rd Edition

It is almost eight years since the first edition of this book was published by NACE International. Since then the demand for corrosion management expertise and services (i.e., training, auditing, consultancy, and field implementation) has been constantly growing, both within the hydrocarbon and the non-hydrocarbon industries.

As an indication, the number of corrosion management training courses I did between 2015 and 2018 had more than quadrupled, while the number of course attendees grew more than tenfold. Simultaneously, an ever increasing number of operator or asset owner companies have turned to the corrosion management concept to further enhance and enrich their existing asset integrity management systems, which were originally based mainly on corrosion engineering approaches and principles.

Such increasing demand–in conjunction with the continuous stream of positive feedback received regarding the first and second editions of this book–made me determined to compile the third edition. This latest edition is almost a total revamp of the previous one and greatly benefits from my continuous learning process through my frequent involvement with various oil and gas sectors and different industries as a corrosion management trainer, auditor, and consultant.

Since the latter half of 2015, when I started working as an independent corrosion engineering and corrosion management consultant, I have been increasingly engaged with numerous upstream-, midstream-, and downstream oil and gas assets as well as in various non-hydrocarbon industries. Such continuous encounters has taught me many new valuable lessons and provided me with a much deeper insight into how we often manage our assets' integrity, our strengths and weaknesses, the sometimes very trivial causes of our integrity oversights and mistakes, and so on. I have done my best to reflect all such new learning and improved asset integrity management insight in this book.

Simultaneously, I have to highlight the following caveat with regard to the corrosion management concept and its implementation process:

Corrosion management is a rather new concept whose roots go back to the U.K.'s offshore oil and gas industry in the early 2000s. Because of such newness, there is unfortunately a great deal of confusion associated with this concept. For example, to many individuals, corrosion management is still a mere synonym for the corrosion engineering concept and discipline, while many others limit its applications to an inspection activity or campaign that is risk-based.

Such misconceptions or misunderstandings–either unintentional or intentional, along with the growing demand for corrosion management services–has unfortunately resulted in situations where some unaware or unscrupulous individuals or organizations offer, market, and conduct so-called corrosion management services (e.g., training, consultancy, and field implementation), which in the best cases, are either corrosion engineering or risk-based inspection services. However, in some of the worst cases, such "pseudocorrosion" management services have resulted in unjustified and unnecessary increases in integrity and corrosion costs, wasted resources–and sometimes they have even *exacerbated* the existing deterioration rates.

I sincerely hope that the readers find the guidance included in this book clear, useful, and practical.

Preface

Since four years ago, I have had more opportunities to engage with the three hydrocarbon sectors (i.e., upstream, midstream and downstream) as well as various non-hydrocarbon industries–either as a CM trainer, or a consultant, or both.

Thus, the increasing demands for CM knowhow in tandem with my increasing CM experience (both as a trainer and consultant) have been the main inspirations for compiling this third edition. To render the book even more useful and practical, more case studies (Appendix A) have been added to it. Furthermore, CM exercises have been added to the book (Appendix B) and are intended to present the reader with an opportunity to test their corrosion management knowledge and use that knowledge in a more practical and effective manner. The associated answers are provided at the end of the exercise questions.

Acknowledgments

I express my deep gratitude to both Mr. Terry Greenfield (NACE International President) and Ms. Jean Broge (NACE International Product Development Manager) for their prompt and kind review and approval of this book's proposal, which enabled me to start the writing process as soon as possible.

I also thank Jean for kindly providing me with another opportunity to work with Ms. Elaine R. Firestone as my book editor. Elaine and I had worked together on my previous NACE book and I consider myself extremely lucky for having been able to work with her again. Elaine has done a superb job editing the original manuscript and rendering it more coherent and understandable. I owe her a great deal for helping me to once again transfer my knowledge and expertise into such articulate language. She is very meticulous, thorough, and professional in her work, and has turned my original work into an organized and eloquent technical book. I do hope that I get the chance to work with her again in the future.

Last but not the least, I am deeply indebted to the following for their continuous support, guidance, and encouragements since I started compiling the first edition of this book in 2010 up to now:

- Dr. Chris Fowler (former NACE International president, United Kingdom)
- Dr. Bijan Kermani (United Kingdom)
- Dr. Torben Lund Skovhus (Denmark)

1 Introduction

1.1 Background

Asset integrity and its management are the indispensable parts of almost all industries, including oil and gas. Corrosion is a major integrity threat for numerous industries, and it adds various (sometimes very significant) costs to an asset's lifecycle, from design to decommissioning. Such costs include

- Design and materials selection
- Chemical treatment
- Inspection
- Corrosion monitoring and fluid sampling
- Coating and cathodic protection systems

In addition to the above categories, a corrosion failure or leak can inflict more costs, such as

- Wasted product costs
- Asset shutdown (or deferred production) costs
- Repair and replacement costs
- Labor and logistics costs

Perhaps nothing better highlights and demonstrates the effect of corrosion on various industrial sectors than the below excerpts from a study carried out by the U.S. Federal Highway Administration:[1]

- The annual direct cost of metallic corrosion in the United States was $276 billion.

- This figure represented 3.1% of the U.S. Gross Domestic Product and was equal to a direct corrosion cost of just under $1,000 per person per year.
- The total direct and indirect corrosion costs were $552 billon, which was equal to 6% of the gross domestic product.

To combat and control corrosion and its consequences, and to reduce the associated costs, corrosion engineering has been used as the main tool and discipline in the oil and gas industry for a very long time. For many assets, corrosion engineering-based (CE–based) asset integrity management systems (AIMSs) are created to facilitate and promote effective corrosion prevention and mitigation.

Appropriate corrosion engineering input during an asset's design stage can significantly improve the overall asset integrity management after commissioning. Once an asset has been commissioned, such input and applications should be reviewed when and where required to ensure they remain relevant, adequate, and effective. However, as the asset ages, the likely operational and process changes may render the existing corrosion engineering applications *less* relevant, *less* up-to-date, and *less* efficient. In such situations, the asset may begin to suffer from corrosion-related problems of increasing number and severity if such changes and their potential effects on the asset's integrity are not addressed in a timely and proper manner.

However, working with and studying various integrity management projects associated with the oil and gas industry have demonstrated that corrosion engineering by itself cannot effectively and efficiently mitigate corrosion. Numerous observations revealed that many corrosion-related failures and leaks were not at all due to obsolete, inadequate, or erroneous corrosion engineering designs or applications but were promoted, accelerated, and bolstered by some non-CE–based shortcomings. Such observations have also shown that the absence or inadequate presence and application of such non-CE–based factors or parameters—either due to negligence, ignorance, complacency, incompetence, or a combination of them—can adversely influence the overall integrity management of an asset. Interactions with other industries (outside the oil and gas industry) demonstrated the same patterns, issues, and causes.

The following list describes some of the observed cases in which non-CE–based parameters or factors were believed to be the main culprits in the loss of integrity:

- The most prevalent deterioration mechanisms behind most of the observed regular equipment leaks and failures—in an asset or within a particular system—were not identified over long periods after commissioning. The failure to identify them often led to an inability to devise and implement the required remedial, corrective, or rectifying activity (or activities).
- No integrity reviews were carried out after commissioning; hence, changes in the operations and process parameters increased the corrosion threat levels or even created new ones. This situation often led to higher deterioration rates and more frequent leaks, equipment failures, and near misses.

- Lack of appropriate and much-needed databases, registers, procedures, strategies, meeting minutes, and other forms of documentation whose data, information, guidelines, and instructions could have significantly improved the efficiency of an existing AIMS.
- Lack of adequate communication between various individuals or companies who contributed to the overall AIMS of an asset or a cluster of them.
- Lack of adequate training associated not only with corrosion engineering but also with corrosion management.
- Reliance on out-of-date, inaccurate, or irrelevant data and information, which often led to erroneous technical judgments and decisions with seemingly higher cost implications.
- Reliance on a non-risk based–approach to inspection and overall asset integrity management.

With the benefit of hindsight, many of the above issues and shortcomings (which led to various corrosion failures) could have been easily and conveniently prevented and mitigated if a simple but well-defined and multistaged process had been in place. This process would have been composed of the following three stages:

1. Reviewing the required integrity management measures
2. Regularly monitoring their performance
3. Assessing their effectiveness

Assets on which this process had been implemented benefited greatly from significantly better and more efficient asset integrity management, associated with optimized costs.

1.2 Rationale

Beginning with the U.K.'s offshore oil and gas industry in the early 2000s, it became clear that relying on the corrosion engineering discipline as the main tool for corrosion control and mitigation was not the right approach and mentality. Simultaneously, it gradually became evident that some non-CE–based integrity management measures played a very crucial role in a successful and efficient asset corrosion control system. Such integrity management measures included

- Communication
- Competency
- Procedures
- Documentation
- Corrosion control matrices
- Risk-based inspection

Thus, the next integrity management challenge was how to incorporate the existing CE–based measures with the non-CE–based ones, while ensuring their smooth and seamless interaction and functioning. Consequently, different corrosion management models were proposed by various organizations and individuals who were trying to address the aforementioned gap or challenge.

Based on the author's oil and gas experience back in the early 2010s, he proposed a new definition for the corrosion management concept and supplemented it with an implementation procedure. Together, the proposed definition and procedure are referred to as the Morshed Corrosion Management Model (MCMM) in this book.

Over the years, this model has been successfully implemented numerous times within the oil and gas industry and then later in various non-hydrocarbon industries. Therefore, the model is presented herein as it was in the early 2010s; however, it is now further complemented with a newly revised narrative, diagrams, examples, and a new set of exercises.

1.3 Scope

The MCMM presented in this book, along with the associated discussions, guidelines, and principles, can be used or easily applied across the three main sectors of the oil and gas industry:

- Upstream Sector: Offshore and onshore production and processing plants, their associated production flowlines, downhole tubing, storage facilities, water injection systems, and utility systems
- Midstream Sector: All product transfer pipelines–whether subsea, onshore, buried or exposed–and water injection pipelines
- Downstream Sector: All oil and gas refineries, storage facilities, and petrochemical complexes

While the oil and gas industry remains the main focus of this book, it is believed that the corrosion management principles discussed herein can also prove equally useful in other industries where corrosion is also a major integrity issue.

1.4 Main Objectives

As has already been stated, the three main attributes of clarity, usefulness, and practicality have always been the author's mission when compiling corrosion management books. Such attributes are crucial prerequisites for this book to achieve its intended objectives:

- To enable the reader to fully appreciate what exactly the corrosion management concept is and how it differs from corrosion engineering and risk-based inspection
- To enable the reader to fully understand the concept of an integrity management measure and how comprehending this concept can revolutionize the way an organization conducts its asset integrity management

- To enable individuals to plan, design, and utilize various corrosion management tools such as corrosion control matrices, corrosion key performance indicators, risk-based inspections, corrosion rate monitoring and fluid sampling, anomalies, and communication
- To guide individuals to have a better systematic approach in identifying integrity issues and then finding or determining solutions, mitigations, or rectification activities.
- To highlight the significance of (the 12) Management Requirements (Chapter 4)
- To assign equal importance (if not more) to the non-CE–based integrity management measures as compared with the CE–based ones.

1.5 Intended Audience

There are no technical prerequisites for using this book so anyone with a basic appreciation of corrosion and asset integrity management can take advantage of the guidance herein. The following personnel are deemed to benefit the most from this book:

- Corrosion and integrity engineers and managers
- Inspection engineers
- Operation engineers and managers
- Maintenance engineers
- Asset project engineers and managers
- Cathodic protection and coating engineers and technicians
- Technical assistants
- Production chemists or chemical treatment specialists
- Process engineers
- Health and Safety executive (HSE) engineers
- Pipeline integrity engineers
- Corrosion monitoring specialists
- Any discipline or individual concerned with asset integrity management, inspection, maintenance, operation, and safety

1.6 Understanding the Terminology

It is expected that corrosion and integrity engineers are the primary personnel interested in this book. Nevertheless, this book is also intended for personnel in other disciplines who often assist corrosion engineers in mitigating corrosion and managing the integrity of the asset. Such disciplines include

- Integrity project engineers and managers
- Chemical suppliers and lab technicians
- Maintenance engineers
- Microbiologists
- Reliability engineers

- Operations engineers
- Risk-based inspection (RBI) analysis (software) technicians and engineers

Because of such diversity, some of the aforementioned disciplines may not be fully and adequately conversant with some of the corrosion management related terminology used in this book. Therefore, Section 1.7 intends to provide clear and detailed definitions for some of the main terms and concepts used.

Of high importance, is the definition offered for the term "risk-based inspection" or "RBI" hereinafter. RBI is probably the most common term used in the field of (pressure systems) integrity management. In spite of its wide use, there appears to be a lack of consensus about its meaning and applications because different users can mean different things when using this term. This can lead to further confusion and misunderstanding–in particular, when negotiating and signing integrity management contracts for pressure systems and equipment.

Another good example is the term "corrosion management" itself, which is the main theme of this book. Unfortunately, to many corrosion engineers, this term is still regarded as a synonym for corrosion engineering. However, corrosion management should be regarded as an independent concept with its own definition, scope, and applications.

Note that all the terms used in this book are based on the definitions in Section 1.7, unless otherwise stated. Furthermore, Section 1.7 includes some terms that are believed to be misnomers and they are *not* used in this book at all. They are nevertheless included here to discourage their further use among other users. A good example is the term "corrosion risk assessment" (CRA), which is used by some as a synonym for "failure risk assessment" (FRA), but it actually conveys the wrong message from a language point of view, as explained in Section 1.7.

1.7 Definitions

Anomaly: An anomaly is a physical characteristic of an item under inspection that detracts from its original design specification. It is considered to be outside certain predefined limits.

Asset: An asset can be an offshore platform, a transfer pipeline, an onshore processing facility, a refinery, or a petrochemical complex.

Asset Corrosion Management Strategy (Asset CMS): An Asset CMS document is a suite of asset-based information and instructions in the form of various corrosion-related databases, drawings, tables, etc., which would help project corrosion engineers to significantly improve their understanding of the asset (i.e., process, operations, design, materials, risk assessment, susceptible areas, etc.) and enable them to make better corrosion engineering and corrosion management judgments and decisions. More specifically, this document can help them to devise and implement enhanced inspection, monitoring, and mitigation activities.

Asset Integrity Management System (AIMS): AIMS is used to refer to the combination of corrosion engineering and non-CE–based applications that are designed and implemented to maintain and/or improve an asset's integrity. Because the scope of this work only covers an asset's pressure systems, the term AIMS covers *only* the pressure systems and excludes other concepts such as structural integrity management.

Criteria of Non-Conformance (CNC): In the context of this book, CNC is often a wall thickness limit, below which the measured wall thickness should be regarded as an anomaly. There are various arbitrary ways for setting the CNC level including the following formulas:

CNC = Nominal Thickness – (Corrosion Allowance + Tolerance), or

CNC = Minimum Allowable Wall Thickness (MAWT)+ 1.0 mm, or

CNC = MAWT + 2.0 mm

Corrosion: The general use of the term "corrosion" throughout this book is intended to cover the other two major deterioration mechanisms—erosion and cracking—unless stated otherwise.

Corrosion Control: This is a general term often used when the intention is to refer to an activity or a collection of activities which when implemented would reduce the amount of corrosion. It is often used to refer to either corrosion engineering or corrosion management or both. Such usage is not advised when either corrosion engineering or corrosion management applications are intended. If the intention is to refer to both corrosion engineering and corrosion management concepts together, then it is better to use the term "integrity management" or "asset integrity management."

Corrosion Control Matrices (CCM): A collection of activities, which if carried out regularly or when required, would maintain an asset's integrity. Typical CCM activities are either monitoring-based activities (both corrosion rate and fluid monitoring) or are of a mitigatory nature, such as corrosion inhibitor or biocide injections. They are determined following the integrity review process and are considered as one of the main products of this process.

Corrosion Engineering (CE): This can be defined as the application of design, material selection, and environmental control principles to prevent or mitigate corrosion. CE is closely associated with an asset's design stage and would continue into the post-commissioning stage where it interacts with the non-CE–based measures. Both CE and CM form the main two components constituting the asset integrity management or the integrity management concept (for pressure systems).

Corrosion Key Performance Indicators (Corrosion KPIs): These are activities that are chosen from the corrosion control matrices document and are deemed to be the most crucial ones in maintaining the asset integrity. They are carried out regularly or when required. Corrosion KPIs are calculated and reported monthly on a percentage compliance basis.

Corrosion Management (CM): This is defined as the process of reviewing the required integrity management measures, the regular monitoring of their performance, and the assessing of their effectiveness post-commissioning.

Corrosion-Risk Assessment (CRA): CRA literally means "assessing the risk of corrosion," whereas the actual and correct process should be to "assess the risk of failure due to corrosion" (see Failure Risk Assessment [FRA] below). Almost all alloys suffer from corrosion to a lesser or greater extent, but the corrosion engineer's interest is when such corrosion or deterioration would lead to failure, hence the term "failure risk assessment." Some corrosion engineers use the term CRA to refer to the failure risk assessment (FRA) process. This is an erroneous practice and CRA should not be used instead of, or as a synonym for, FRA.

Failure Risk Assessment (FRA): FRA is the process through which the risk of equipment failure due to corrosion, erosion, and cracking is assessed and failure risk rankings, and inspection intervals are assigned to the equipment being assessed. This process is often and erroneously referred to as "risk-based inspection" (RBI), which is a misnomer and should be avoided. The other term used to refer to the FRA process is corrosion-risk assessment (CRA, see above), which should be avoided as well.

Integrity: The ability of an asset (or parts of it) to function as per its intended design specifications in processing and containing produced hydrocarbons in a safe manner until the next planned inspection or scheduled repair and replacement.

Integrity Management (IM): Maintaining and/or improving the integrity of an asset through corrosion engineering and corrosion management practices and applications.

Integrity (Management) Measure: Anything (e.g., an activity, document, procedure, and an individual's competence) that contributes to an AIMS and influences its functioning, effectiveness and outcome.

Integrity Review Process: The process of reviewing the required CE-based and non-CE-based integrity management measures post-commissioning. It can be carried out by a corrosion engineer or a group of people, each one representing a discipline, depending on project requirements and resources. It can be done on a qualitative basis (when no detailed information is available) or on a fully quantitative basis when both information/data and resources are available.

Pressure Systems: The term "pressure systems" in this book refers to pressure piping (both process and utility), pressure vessels, heat exchangers, tanks (including atmospheric tanks), pressure safety valves, and pipelines.

Risk-Based Inspection (RBI): RBI refers to an inspection type where the criterion of risk has been used to create the inspection scope. Other inspection types use other criteria such as time, condition, and consequence to determine the inspection scope (see Chapter 6 for more details).

2 Integrity, Corrosion, and Corrosion Engineering in the Oil and Gas Industry

2.1 What is Integrity?

In general use, the word integrity means "completeness" or "oneness." In an industrial or technical context, however, it means the equipment's ability to function in a safe manner while remaining mechanically fit. In the oil and gas industries, having integrity enables a piece of equipment to function and operate properly so that, for example, the equipment's (hydrocarbon) fluids are kept inside at all times.

Consequently, the term "integrity management" can be regarded as a continuous process throughout the asset lifecycle. This ensures that the physical or mechanical condition of the asset is maintained in such a way that there are no leaks, and all the asset equipment operate and function in a safe manner per their original design specifications or requirements.

2.2 Integrity Threat Categorization

Throughout the asset lifecycle, an asset owner or operator has to deal with various integrity threats. Such integrity threats (Figure 2.1) vary in nature and origin:

- Natural disasters
- Poor operations
- Enemy activity
- Effects of the external environment and the internal process fluids (i.e., corrosion, erosion, and cracking)
- Miscellaneous (e.g., design and mechanical causes)

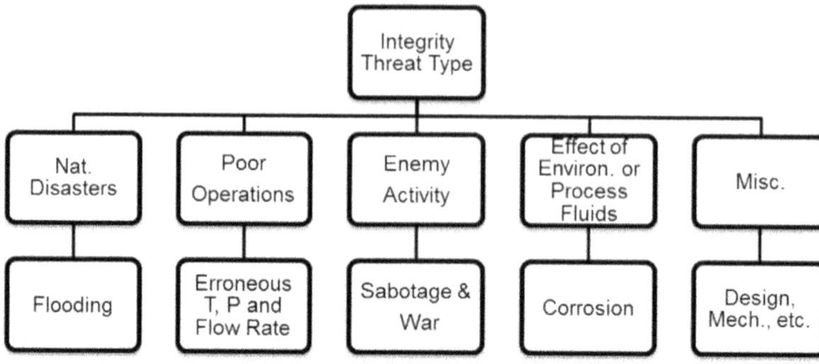

Figure 2.1 Integrity threat types and the associated examples.

To control, prevent, or mitigate such integrity threats for any particular asset or a collection of assets, there must exist an asset integrity management system (AIMS) where all such threats have already been taken into consideration, assessed, analyzed, and quantified. In addition, the associated preventive, mitigation, and remedial activities must be clearly described and devised in detail. This AIMS is a live document that should be reviewed, revised, and updated on a regular basis–where and when required.

However, throughout this book the main focus will be on the effects of the external environment and the internal process fluids that often manifest themselves as corrosion, erosion, and cracking. Thus, the other integrity threat types are not covered in this book.

2.3 Corrosion in the Oil and Gas Industry

Corrosion is considered a major integrity threat for many different assets in the oil and gas industry. Corrosion accidents, leaks, and failures cause frequent unplanned asset operation shutdowns and inflict various costs:

- Wasted product costs (where applicable)
- Repair and replacement costs
- Deferred production costs (where applicable)
- Logistics and labor costs
- Asset downtime costs
- Legal costs and fines

Corrosion engineering (i.e., design, materials selection, and environmental control) had been used extensively and progressively to mitigate corrosion and its associated varied costs and repercussions. For almost a century and up to the early 2000s, corrosion control and mitigation had been carried out almost entirely through corrosion engineering. This was because design, materials selection, and environmental control were considered as the main components of the corrosion engineering discipline or concept. The view, mentality, or approach that corrosion engineering by itself can effectively and thoroughly mitigate corrosion throughout the asset lifecycle is referred to as the Classical Integrity Management View, which is discussed in detail in the next section.

2.4 The Classical Integrity Management View

This view is a notion whereby corrosion control or mitigation is mainly implemented via the corrosion engineering discipline. The Classical Integrity Management View persisted almost a century, until it was challenged by the integrity findings and observations pertaining to the U.K.'s offshore oil and gas industry in the mid-1990s onwards[2-3] and then supported by similar findings internationally.[4]

These integrity findings indicated that the corrosion engineering discipline was not able to both effectively and efficiently mitigate corrosion by itself and there were various other parameters and factors (or more precisely there were other critical integrity management measures) that simultaneously played a crucial role in maintaining asset integrity and efficiently mitigating its deterioration, as detailed in Section 2.6.

2.5 The Corrosion Engineering Concept

Corrosion engineering can be defined as the design and application of methods to prevent corrosion. That is, any factor or parameter that directly affects the corrosion rate or the corrosivity of the environment can be considered as a component of corrosion engineering. As previously stated, the main components of corrosion engineering are

- Design (of assets and equipment)
- Materials selection
- Controlling the environment (e.g., chemical treatment)

Figure 2.2 illustrates the three main components of the corrosion engineering discipline and their associated subcomponents.

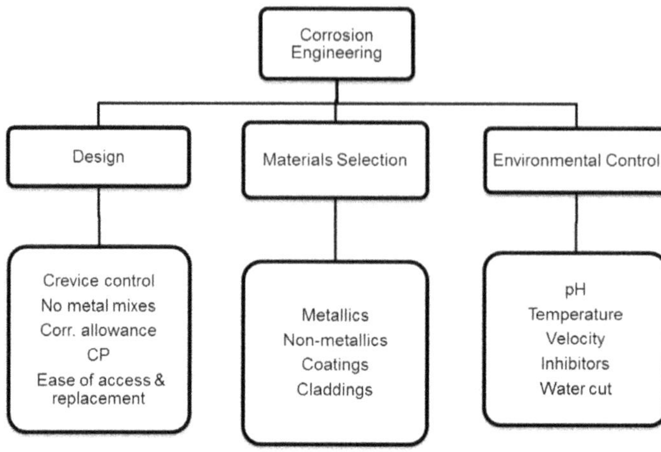

Figure 2.2 The corrosion engineering discipline, its three components, and subcomponents.

2.6 Shortcomings Associated with the Classical Integrity Management View

As mentioned in Section 2.4, the U.K.'s offshore oil and gas industry experience and observations had shown that there were many other factors and variables that significantly contributed to or affected corrosion, its severity and extent, detection, analysis, treatment, and eventual mitigation. None of these factors, however, were associated with the corrosion engineering discipline, its various components, or applications:

- Strategies
- Procedures
- Management of corrosion monitoring and fluid sampling systems
- Management of risk-based inspections
- Documentation
- Communication
- Competencies
- Team structure, roles, and responsibilities

The author's U.K. field experience had also shown that such factors and variables–however simple, commonplace, and ordinary as they may seem–did play a significant role in the overall integrity management of an asset where corrosion was one of the main integrity threats.

The author's international (integrity management) experience within the oil and gas industry in general proved the same thing, i.e., corrosion engineering had to be supplemented with various non-corrosion engineering factors or measures to render corrosion control and mitigation both more effective and efficient.[4] Simultaneously, interactions with his counterparts in other industries pointed to the same culprits and root causes. It became obvious that in other industries–similar to the oil and gas industry–corrosion engineering by itself can not successfully control and mitigate corrosion.

Various field observations (associated with the U.K.'s offshore oil and gas industry) also illustrated that in spite of an initial appropriate corrosion engineering design and associated applications, various equipment eventually leaked or failed because of corrosion–sometimes shortly after their commissioning. This was because some of the non-corrosion engineering factors or parameters were ignored or implemented inadequately, wrongly, or did not exist at all throughout the operation or the commissioning phase of an asset's lifecycle. Consequently, proper introduction, implementation, reviewing, monitoring, and assessing of such factors and variables can significantly enhance AIM through improving corrosion mitigation and control.

Based on what has so far been mentioned, the shortcoming of the Classical Integrity Management View was that it had erroneously presumed for almost a century that corrosion engineering by itself can adequately and efficiently control and mitigate corrosion in the oil and gas industry. That is, design, materials selection, and environmental control were the only factors or more precisely–the only integrity management measures–that can influence an asset's integrity and control, and mitigate its corrosion. The above erroneous notion manifested itself in three different asset integrity assessment stages as depicted in Figure 2.3.

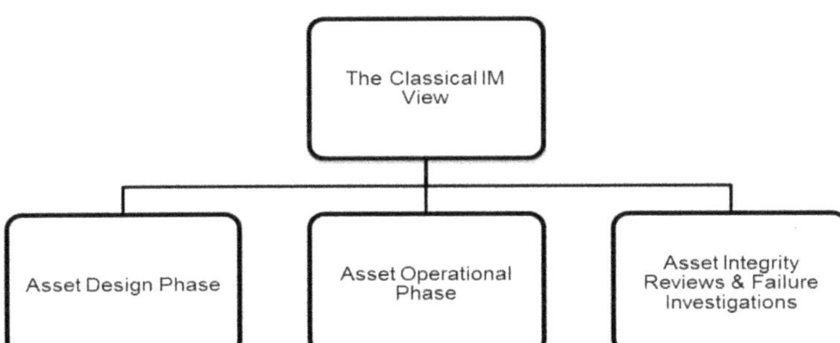

Figure 2.3 The three different asset integrity assessment stages where only the CE-based parameters were considered.

During the design considerations and review, the operational phase, and when conducting asset integrity reviews and failure investigations, only the corrosion engineering components were considered and the non-CE-based parameters (such as communication, competency, corrosion control matrices, and procedures) were completely or partially excluded. Such an approach over time resulted in various corrosion and integrity failures, issues, and shortcomings.

As such shortcomings gradually became evident, efforts began by both some organizations and individuals to introduce a model or propose a more comprehensive integrity management view to compensate for such a deficiency and also to incorporate the non-CE–based factors and parameters into any such new model. Such efforts over the years has led to the creation of two newer AIM notions, better known as *corrosion management models*, which are discussed in Sections 2.7.1–2.7.2.

2.7 The Two Current Corrosion Management Models

Since the early 2000s, continuous efforts to offer a more comprehensive AIM approach or view (as compared with the then existing and defective Classical Integrity Management View) has so far resulted in two internationally recognized asset corrosion management models.

The first one has been developed and promoted through the continuous support and endorsements of the U.K.'s Health and Safety Executive (HSE) organization since the early 2000s, and is known as the "UK HSE Corrosion Management Model."

The second one was created and offered by the author in the early 2010s, and is known as the "Morshed Corrosion Management Model" (hereafter, MCMM). Interestingly, the origin of both of these asset corrosion management models was the U.K.'s offshore oil and gas industry.

The following two sections briefly introduce each model and discuss their pertinent attributes.

2.7.1 The U.K. Health and Safety Executive (HSE) Asset Corrosion Management Model

The advent of this model goes back to 2001 when one of the first HSE-sponsored corrosion management reports was published.[2] The corrosion management model presented therein was further revised when the U.K. Energy Institute published upgrades of the original model in May 2008,[5] and March 2019.[6] There are four main attributes of the model presented in the 2008 publication:

1. It is mainly intended for the upstream sector of the oil and gas industry (Ref. 5, Section 1.2.2).
2. Downstream hydrocarbon industries and non-hydrocarbon industries are excluded (Ref. 5, Section 1.2.2).
3. It mainly focuses on CE–based deterioration factors (Ref. 5, Section 1.2.3).
4. This model regards corrosion management as a synonym for corrosion engineering (Ref. 5, pages 23–24).

2.7.2 The Morshed Corrosion Management Model (MCMM)

Because the MCMM is the main focus of this book and is covered in detail in the following chapters, it is introduced very briefly in this section. The origin of this model goes back to the early 2000s when the author started working first for the U.K.'s offshore oil and gas industry (based in Aberdeen), and then after a few years began working internationally.

In the mid-2000s he started working on a new asset integrity management (AIM) approach in which he strived to eliminate the existing deficiencies associated with the then-predominant view that corrosion engineering was the sole solution for corrosion issues in the upstream oil and gas industry in the United Kingdom. Later, he came up with a process that was composed of three smaller processes and called it the "corrosion management process" or simply "corrosion management." His new AIM approach was based on the fact that two different types of integrity management measures played critical roles within any AIMS. The first type or group was the CE-based integrity management measures, and the second was the non-CE-based integrity management measures. Since 2007, his new approach has been covered and promoted through numerous NACE International publications (both as papers and books).

There are six main attributes of the MCMM:

1. It offers an easy, clear, and practical definition for the term "corrosion management."
2. It offers the new concept of the "integrity management measure" to enable the model to capture and cover both the CE-based and the non-CE-based factors and parameters (which play a crucial role in any AIMS).
3. The model is applicable to upstream, midstream, and downstream oil and gas sectors.
4. It is also applicable to non-hydrocarbon industries.
5. The model highlights and promotes the significance of non-CE-based integrity management measures.
6. It further emphasizes the criticality of communication and competency as the most significant non-CE-based integrity management measures.

2.8 Conclusions

Corrosion is an integrity threat, to various extents, for the majority of oil and gas assets around the world. The notion that the corrosion engineering discipline (through its three components of design, materials selection, and controlling the environment) can efficiently control and mitigate corrosion throughout an asset's lifecycle is referred to as the "Classical Integrity Management View." However, since the early 2000s, and beginning with the U.K.'s offshore oil and gas industry, it gradually became evident that the aforementioned view was deficient and that there were also other non-CE–based factors or parameters that also played a crucial role in any AIMS. Consequently, efforts began to address such inherent shortcomings and they have so far led to the creation of two internationally better-known corrosion management models.

3 The Corrosion Management Concept and Its Implementation

3.1 Background

The inherent deficiencies associated with the Classical Integrity Management View and its failure to efficiently and successfully control corrosion in the U.K.'s offshore oil and gas industry encouraged various organizations and individuals to try to come up with an enhanced and more effective asset integrity management approach or view (see Section 2.6). To address the then-existing asset integrity management (AIM) shortcomings, in the early 2010s the author proposed a new integrity management approach through a newly defined process, and simultaneously proposed a new definition for this new process, which he called the "corrosion management process" or "concept". This chapter discusses the origins, concept, components, and finally the implementation process associated with this model.

3.2 The Missing Links or Processes

Beginning with the U.K.'s offshore oil and gas industry throughout the early 2000s, it gradually became clear for the author that what distinguished successful asset owners or operators from the less successful ones–in the area of asset corrosion control and mitigation–was adhering to, or carrying out, three simple processes:

1. Integrity review process
2. Performance monitoring process
3. Effectiveness assessment process

Implementing such simple and straightforward processes–for both CE-based and non-CE-based parameters or factors–made a significant improvement in the way the integrity of an asset was being managed. Similar observations were made several years later when the author had the opportunity to work on various international assignments in different parts of the world such as Eastern Europe, North Africa, the Persian Gulf Region, and South Asia. The associated observations and the ensuing conclusions were exactly the same: that when the above three processes or links were present in a timely and adequate manner, they not only enhanced the corrosion failure preemption capability of an (oil and gas producing and processing) organization, but also simultaneously optimized its corrosion and integrity costs.

The observations associated with these three processes or links are discussed in the following three sections along with some pertinent examples.

3.3 Missing Link 1: The Integrity Review Process

Reviewing any variable or factor that affects the overall functioning and output of an integrity management measure (either CE-based or non-CE-based), whether it has already been implemented or is to be devised, is called the Integrity Review Process (IRP). During this process the relevance, applicability, effectiveness, justifications, and possible outcomes of that particular integrity management measure on the incumbent (or the future) asset integrity management system (AIMS) is reviewed. Simultaneously, all the relevant factors, parameters, components, or variables that control such measures and their effective and successful implementation are also reviewed.

In spite of the significance of such a crucial and indispensable process, unfortunately many assets lacked a proper and adequate integrity review process to cover both CE-based and non-CE-based variables. The direct effect of such negligence or complacency is the gradual increase in both corrosion rates and the number of corrosion failures and leaks. The following integrity review examples will make comprehending this vital process easier:

- **Corrosion Inhibitor Injection Integrity Review:** For an effective and successful corrosion inhibitor treatment, one should carefully review and consider as many of the following pertaining variables as possible:
  - Whether corrosion inhibitor injection is at all required or not (for example, the water cut might be too low, and considering the flow conditions within the transfer pipelines such a chemical treatment may not be justifiable)
  - The type of chemical required (often determined by lab testing)
  - The stability conditions (for example the environmental pH and temperature, which affect the performance and stability of such chemicals)
  - The required chemical concentration (often determined by lab testing)
  - The injection location
  - The injection method (e.g., the necessity to use injection quills for better mixing of the injected chemical with the process fluids)
  - Continuous or intermittent treatment

- The order of injection (in cases where more than one chemical is injected into the system)
- Compatibility of the injected chemical with other metallic and nonmetallic materials already present in the system
- Weld protection (i.e., whether the protection offered by the chemical to the welds is as good as, or as near as, the protection offered to the other parts of the system)
- Sand protection (whether such a chemical treatment is effective in systems where the lower parts of the system are covered by the produced sand)

The author has come across numerous cases where either one or several of the above injection variables had not been reviewed for many years and the chemical treatment eventually became totally obsolete. In one particular case, where the maximum recommended injection rate (or concentration) was 20 ppm, the operator had been injecting the chemical at 3,000–4,000 ppm for a prolonged period until the integrity review identified such unbelievable chemical overdose.

In another case, high corrosion rates had been detected just upstream of the corrosion inhibitor injection point. The integrity review later revealed that the location had to be moved farther upstream to protect those high corrosion rate areas. Such a move later rectified the situation.

- **The Process Equipment Inspection Integrity Review:** Ideally the ultrasonic testing (UT) wall thickness inspections carried out in the oil and gas industry should be fully risk-based. Numerous site visits have revealed that many inspection plans, despite claims by the incumbent inspection company or contractor, were not risk-based at all. In other cases, integrity reviews showed that the wrong inspection technique had been used—the data analysis techniques were insufficient or erroneous and that the staff had not been adequately trained for the job, if at all. In general, various factors affect the outcome of an inspection campaign:
  - Inspection technique (e.g., UT wall thickness inspection, radiography, visual, and eddy current)
  - Inspection type (e.g., time-based, condition-based, or risk-based)
  - Competency of the inspection team
  - Implementation of the inspection technique (e.g., whether or not any required equipment calibration takes place)
  - Inspection frequency
  - Inspection extent

- **The Communication Integrity Review:** Repeated site visits and the ensuing integrity reviews had revealed that the principal culprit behind the incumbent high corrosion rates and frequent operation shutdowns was the total lack, or inadequate levels, of communication within the asset integrity management (AIM) team. In fact, poor communication has proven to be one of the main shortcomings exacerbating equipment deterioration due to corrosion worldwide. Therefore, reviewing the following variables that affect the quality and effectiveness of any communication are very important:

- Who reports to whom
- Frequency of reporting or communication
- Parameters or data types included in the report or communication
- The report template (e.g., the specific word processing or presentation software used)
- Manner of communication (e.g., meetings, online, written, or teleconference)

- **The Competency Integrity Review:** Lack of, or inadequate, competency levels are behind many integrity and corrosion incidents. Incompetent engineers (such as corrosion, inspection, process, and pipeline engineers) contribute significantly to assets and equipment deterioration. Any integrity reviews carried out should also take into consideration the competencies of the key people or disciplines within the asset integrity management team. Competency parameters that should be taken into consideration during any integrity review process should include
    - Education
    - Experience
    - Training
    - Supervision

- **Coating Selection and Application Integrity Review:** For any coating to perform satisfactorily it has to be selected and then applied properly. Parameters that should be taken into consideration during an integrity review when selecting and applying a coating include
    - Type of the substrate
    - Environmental conditions (once the coating has been applied)
    - Surface preparation requirements
    - Number of coating layers to be applied
    - Application type
    - Application conditions (when the coating is being applied)
    - Lifetime requirements
    - Maintainability
    - Dried film thickness

3.4 Missing Link 2: The Performance Monitoring Process

To efficiently mitigate corrosion, a combination of CE-based and non-CE-based variables and parameters have to be present and implemented. Thereafter, one has to devise a system or methodology to ensure the associated performances (of both types) are constantly being monitored. However, in many of the visited sites, or audited and studied AIMSs, the performances of neither the CE-based parameters nor the non-CE-based ones were being monitored. That is, nobody had any ideas what exactly the performances of such parameters were. Such a critical shortcoming further exacerbated the implementation of the third process, which was the effectiveness assessment process. Conspicuously, because there were no performance figures, their effectiveness could not be assessed or evaluated.

The following are just a few examples pertaining to the lack of the performance monitoring process:

- Performance monitoring process associated with corrosion inhibitor injection: A chemical has been injected for several years into the system as corrosion inhibitor. However, the corrosion rate downstream of the injection point has never been measured. Therefore, no one can assess the effectiveness of such chemical treatment to determine whether it can be continued or if it has to be replaced.
- Performance monitoring process associated with a typical internal coating: A coating has been applied on the internals of various pressure vessels and is intended to last for a prolonged period of time. However, during numerous operation shutdowns and internal inspections of the those vessels, the associated coating performance has never been measured or evaluated in anyway, in spite of coating defects and deficiencies observed during each shutdown. Therefore, it is not possible to determine whether it has been an effective coating or has to be replaced or upgraded.
- Performance monitoring process associated with competency levels: A corrosion engineer has been acting as the sole asset corrosion engineer for a number of years; however, this engineer's competency (levels of education, experience, training, and supervision) has never been measured or evaluated. Therefore, no one can determine whether the engineer has been doing a good job or not.

3.5 Missing Link 3: The Effectiveness Assessing Process

Once the performance of a particular parameter has been measured, then one has to decide whether such performance has been acceptable or not. To do that, the measured (performance) figure has to be compared with a well-defined limit, figure, or range to determine its effectiveness. The following are just two examples:

- Effectiveness assessment process for corrosion inhibitor injection: The inhibited corrosion rates downstream of the injection point have to be measured and then compared with a well-defined limit or range to determine whether the corrosion inhibitor treatment has been effective enough to reduce the uninhibited corrosion rates below an acceptable limit or not. If the answer is "no" then one has to carry out an integrity review process on the chemical injection to identify the associated shortcomings or deficiencies, and then carry out relevant activities to ensure that the inhibited corrosion rates fall below the defined or agreed-upon corrosion rate limit.

- Effectiveness assessment process for a cathodic protection system: A cathodic protection system provides protection against external corrosion for a buried pipelines system. Once the pipeline potential has been measured at various locations, the measured values have to be compared with an acceptable or defined potential range to determine the effectiveness of the cathodic protection system. If the measured figures are not within the acceptable range, then an integrity review process has to be carried out to identify the "culprits," i.e., what is preventing the cathodic protection system to function properly.

Therefore, if the effectiveness assessment process is not carried out (either because its prerequisite process–i.e., the performance monitoring process–had not been carried out in the first place, or for any other reason) there is no way to determine whether an incumbent integrity measure or variable has been performing satisfactorily or not. That is, its effectiveness would remain unknown.

Unfortunately, not knowing the level of effectiveness has been a common observation in many of the oil and gas sites the author visited. In such sites the effectiveness assessment process had not been carried out for the pertaining integrity management measures, and some of them, which had been ineffective since the first implementation day, had remained in place in spite of such poor performance.

3.6 The Integrity Management Measure Concept Definition

To be able to carry out the three processes described in Sections 3.3–3.5, one has to initially identify the variables or parameters for which such processes can be carried out or implemented. The selection of such variables necessitates creating a criterion or a set of criteria to enable such selection. Consequently, such criterion can be defined as any variable, factor, or parameter that contributes to an AIMS and influences its functioning, effectiveness, outcome, and successful performance. Integrity management measures are either CE-based or non-CE-based. Examples include

- Materials selection (CE-based)
- Chemical treatment (CE-based)
- Cathodic protection (CE-based)
- Coatings (CE-based)
- Inspections (non-CE-based)
- Registers (non-CE-based)
- Procedures (non-CE-based)
- Documentation (non-CE-based)
- Competency (non-CE-based)
- Communication (non-CE-based)

The integrity management measure concept plays a central and very crucial role in the MCMM presented in this book. The model covers both the CE-based and the non-CE-based integrity management measures as mentioned above. Without the integrity management measure concept, the MCMM would never function or operate properly and adequately; therefore, the significance of this concept has to be fully appreciated prior to discussing the corrosion management definition itself.

It has to be reiterated here that according to the Classical Integrity Management View, only the corrosion engineering three main components (i.e., design, materials selection, and controlling the environment) were considered as integrity management measures and the non-CE-based ones were somehow excluded. This viewpoint constituted the most important deficiency of the above notion and limited it to only CE-based integrity management measures. Thus, the above view was unable to identify, rectify, solve, or improve asset integrity cases where the roots were totally or partly associated with the non-CE-based integrity management measures such as documentation, communication, competency, and inspections.

3.7 The Corrosion Management Concept Definition

Once the integrity management measure concept has been fully introduced, explained, and comprehended, then one can move to the next most important concept, which is the corrosion management concept itself.

According to the MCMM, the corrosion management concept can be defined as

> "The process of reviewing the required integrity management measures, regular monitoring of their performance, and the assessment of their effectiveness post-commissioning."

With regards to the above definition, the following points have to be noted:

- The phrase "all the required integrity management measures" refers to both existing and future CE-based and non-CE-based activities, applications, and requirements. That is, when one is carrying out the integrity review process to identify the incumbent deterioration mechanisms they have to consider the future ones as well, so they can plan in advance and be ready for these mechanisms when they occur.
- The corrosion management process should be applied throughout the asset lifecycle; that is, from design, to construction, to transportation, to installation, to hydrotesting, to operation, and finally to decommissioning. However, because in this book the main emphasis is on corrosion management applications during the operation phase (of the whole asset lifecycle) then the term "post-commissioning" has been added to the definition to highlight such applications.

It has to be highlighted here that the terms "corrosion management" or "asset corrosion management" have been in use for a very long time, mostly as a synonym for the corrosion engineering term. When used in such a context corrosion management often meant corrosion control, corrosion mitigation, or corrosion prevention through the three corrosion engineering components. However, according to the above new definition, "corrosion management" is now the term for a newly introduced process that is composed of three smaller processes. Therefore, throughout this book whenever "corrosion management" is used it refers to the above process composed of integrity review, performance monitoring, and effectiveness assessment processes.

Figure 3.1 is an illustration of the newly defined corrosion management process.

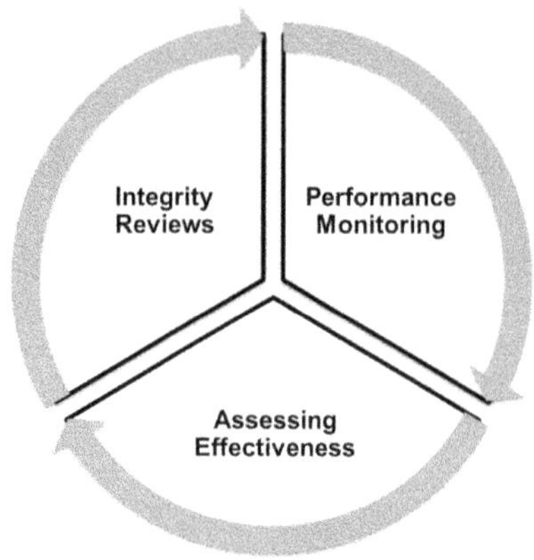

Figure 3.1 The three stages of the corrosion management process or concept.

3.8 The Morshed Corrosion Management Model (MCMM)

The MCMM is composed of two main components:

1. The "corrosion management process," which itself is composed of three smaller processes (introduced in Section 3.7).
2. The "integrity management measure" types chart depicted in Figure 3.2.

Once the above two components are put together, then the (three-staged) corrosion management process should be applied on both the CE–based and the non-CE–based integrity management measures.

According to Figure 3.2, there are three CE–based integrity management measures:

- Design requirements (e.g., determining the shape, geometry, and corrosion allowance of an item)
- Materials selection requirements (e.g., determining whether an item is made from a metallic or a nonmetallic material and then specifying its type, grade, etc.)

- Environmental control requirements (e.g., controlling the pH, scaling tendency, or corrosivity of the environment through injecting a chemical)

In addition, there are 15 non-CE-based integrity management measures:

- 12 management requirements
- Inspection requirements
- Corrosion rate monitoring and fluid sampling requirements
- The failure risk assessment requirements

It is very important that the (three-staged) process defined in Section 3.7 is carried out for both integrity management measure types. This is considered as one of the strengths of this model that assigns equal importance to both types; unlike some previous notions that only considered the CE-based ones. In general, the following are the main attributes of this particular corrosion management model:

- Easy, clear, and practical (corrosion management) definition
- Incorporating the integrity management measure concept
- Applicable to upstream, midstream, and downstream hydrocarbon sectors
- Applicable to non-hydrocarbon industries as well
- Highlighting the significance of the non-CE-based integrity management measures

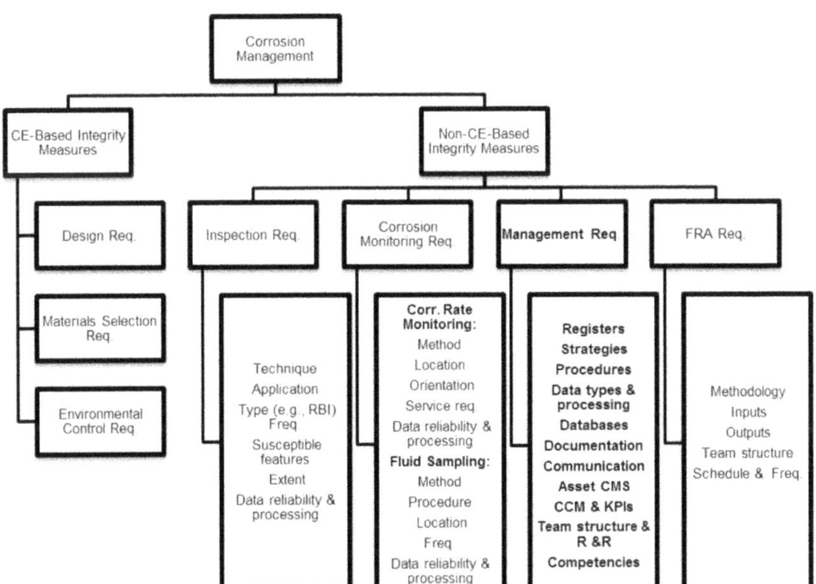

Figure 3.2 The two integrity management measure types covered in the corrosion management process: CE-based and non-CE-based.

3.9 Corrosion Management Misconceptions and Abuses

Because of its relative newness, the "corrosion management" concept and term is still sometimes mistaken for other processes or concepts such as the "corrosion engineering" concept. Unfortunately to many individuals and organizations around the world, the term "corrosion management" is still a mere synonym for "corrosion engineering." Therefore, many of the books, training courses, and consultancy services that are being offered as corrosion management products and services are actually corrosion engineering ones at best. This misunderstanding and confusion is sometimes inadvertent, but in many occasions they are intentional and designed to take advantage of the constantly growing demand for corrosion management services, along with the relative unawareness (about the corrosion management concept) among some asset owners and operators.

Such growth for corrosion management services has given rise to some unscrupulous individuals and organizations selling some pseudo-technical services under the name of "corrosion management" products and services. What they are actually providing is neither corrosion management, nor is it beneficial to the asset concerned from an integrity point of view. The author has unfortunately come across many such unprofessional and unethical practices where the asset owners or operators were deceived and the so-called "corrosion management" services they received actually exacerbated the integrity condition of the asset.

In general, the following misconceptions and confusions still exist in relation to the corrosion management concept:

- FALSE: Corrosion management is a synonym for the "corrosion engineering" term and means reducing the incumbent corrosion rate through design, materials selection, cathodic protection, coating applications, chemical treatments, and pH control.
- FALSE: Corrosion management is identical to the "risk-based inspection" (RBI) process. That is, if in any asset the incumbent inspection scope is risk-based then corrosion management has been implemented therein.
- FALSE: Corrosion management is systematic reduction of corrosion and integrity costs. That is, any cost cutting–even if it increases the corrosion failure risk–is considered as the implementation of the corrosion management process.

Therefore, those asset owners or operators who seek corrosion management services should be conversant with the definition of the corrosion management concept offered herein. Such familiarity with the concept and the corrosion management process will minimize such potential confusions and abuses.

3.10 Introduction to the Integrity Review Process

All corrosion management implementation projects begin with the integrity review process, which is considered as the most crucial process among the three processes (Section 3.7) that comprise the corrosion management process.

It is during this process that a decision is made whether a new integrity management measure needs to be created, or if it already exists, whether it is still effective, relevant, up-to-date, practical, and useful. It has to be reiterated that like the other two processes (i.e., performance monitoring and effectiveness assessment), the integrity review process is applied on both the CE-based and the non-CE-based integrity management measures. Figure 3.3 is a chart that can be used when carrying out the integrity review process.

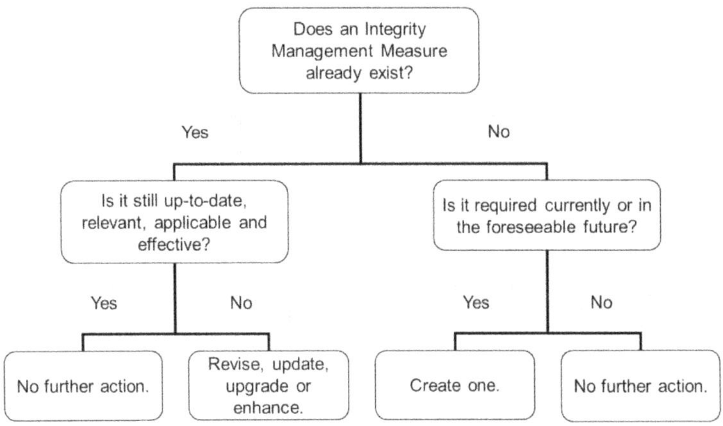

Figure 3.3 The proposed chart for carrying out the integrity review process.

One of the major shortcomings when carrying out the integrity review process is that the review process is often limited to the CE-based integrity management measures and the non-CE-based ones are excluded. Therefore, if the root cause of the incumbent corrosion failure pertains to the latter type (i.e., non-CE-based) then they cannot be identified, addressed, or rectified. Another common shortcoming is that many operators carry out integrity reviews only using inspection data; whereas, there are numerous input data types as illustrated in Figure 3.4.

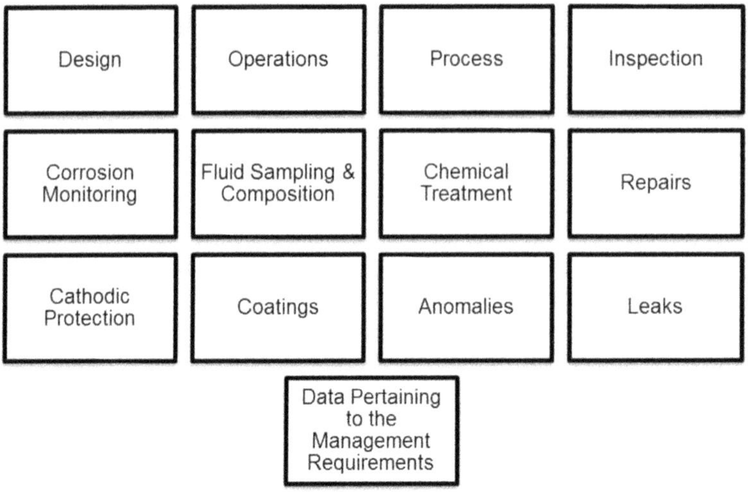

Figure 3.4 Various input data types that should be used during an integrity review process.

Last but not the least, the integrity review process should be carried out to build a picture of the integrity condition of an asset, so one can determine where exactly the integrity issues are and at what intensity, extent, or level. The integrity picture becomes clearer and more complete when more data types and more reliable input data are used. Therefore, the incumbent issues can be identified more easily and more accurately and consequently better integrity fixes or solutions can be devised and implemented for them.

3.11 Introduction to Performance Monitoring Process

Once all the planned or devised integrity management measures (both CE-based and non-CE-based ones) are in place and implemented (if and where possible), their performance should be monitored, recorded, and reported on a regular basis. Such performance monitoring would later enable one to compare the measured performance against an applicable standard, an acceptable value or threshold, and so on. Through such comparison, one can later assess the effectiveness of a particular integrity management measure, such as a specific chemical treatment (i.e., a CE-based integrity management measure), or a bespoke sand policy (i.e., a non-CE-based integrity management measure), or the competence of an asset's corrosion engineer. Then one can decide whether these integrity management measures are working effectively or require a change or improvement to boost their performance.

As mentioned earlier, integrity management measures are either CE-based or non-CE-based and as such, their performance monitoring is quite different from each other.

Performance monitoring for CE-based items (i.e., design, material selection, and environmental control) is often based on the measured, estimated, or modeled annual corrosion rate, or the determined or measured fluid or environment corrosivity. Such measurements come from inspection, corrosion rate monitoring, fluid sampling, and corrosion rate modeling software.

On the other hand, monitoring the performance of non-CE-based items is not usually as straightforward as for those CE-based ones. Performance monitoring for the non-CE-based integrity management measures can be done via counting the number of annual leaks due to corrosion. This is because such integrity management measures, once properly implemented, often do not have a direct effect on the corrosion rate, but after a proper and timely implementation, they can influence the overall AIMS and its effectiveness. In addition, they collectively reduce the incumbent corrosion rates and consequently the number of annual corrosion leaks.

3.12 Introduction to Effectiveness Assessing Process

Once an integrity management measure's performance has been measured and recorded, it should be assessed against a standard or acceptable value or threshold. This is to ascertain whether such performances are acceptable, compliant, and satisfactory. If the measured performance value falls below the acceptable threshold then one has to immediately find out (through carrying out an integrity review process) whether that had been an isolated case or a more frequent one. In the former case, any shortcomings causing such unacceptable performances have to be rectified as quickly as possible. And in the latter case, a review has to be carried out (as soon as such poor performance has been identified) to determine the reason behind such recurring poor performances.

Thereafter, and once the activity or item in question has been amended, revised, or rectified, the performance should rise to levels equal to or above the acceptable thresholds. The following examples are intended to further elucidate how effectiveness is assessed and (if needed) improved.

Example One: Many assets rely on the combination of a carbon steel alloy and a corrosion inhibitor injection for corrosion mitigation during their design life. For assets with a design life of 30 years, the allocated corrosion allowance (for their process piping) is often 3 mm; assuming a maximum annual corrosion rate of 0.1 mm/y. Post-commissioning, if performance monitoring–through inspections and corrosion rate monitoring systems–detected a corrosion rate of greater than 0.1 mm/y, i.e., that the corrosion allowance would be wholly consumed prior to the end of the design life, it can mean one of two things:

1. The initial design assumptions and considerations were erroneous and the carbon steel and corrosion inhibitor combination was not the right choice.

2. The initial design assumptions and considerations were correct at the time, but some process (and/or operations) parameters had varied over time, resulting in more corrosive conditions and higher corrosion rates.

In either case, one has to devise an improved CE-based integrity management measure. Possible improvements include the following measures:

1. Replacing the incumbent chemical with a more effective one
2. Replacing the carbon steel alloy with a corrosion resistant alloy
3. Using a carbon steel grade with a ticker corrosion allowance layer
4. Modifying the process conditions–if possible–to render the environmental conditions less corrosive

Equipment tends to fail with increasing frequency and severity–without anybody bothering to find out the cause or the culprit–wherever no adequate performance monitoring and effectiveness assessment is in effect. This problem often leads to much higher repair-, replacement-, and maintenance costs as well as more frequent operations shutdowns. In such scenarios, real causes of equipment failure (or higher than normal deterioration rates) would remain undiscovered, uninvestigated, and finally untreated for long periods of time; rendering the overall AIM both extremely troublesome and costly.

Example Two: An offshore platform uses a seawater injection system to inject seawater into its subsea wells to maintain production pressure. The concentration of oxygen in the uplifted seawater and upon entry into the seawater injection system is around 5–10 ppm. This has to be reduced to around 10 ppb to render the seawater less corrosive toward the carbon steel piping, farther downstream of the system.

There is an oxygen probe fitted downstream of the deaerator tower and oxygen scavenger injection point. In this setup, a functioning and fully calibrated oxygen probe carries out the (regular) performance monitoring by measuring the residual oxygen concentrations downstream of the scavenging point. Simultaneously, the asset corrosion engineer should have access to the probe data to assess the effectiveness of the oxygen scavenging process on a continuous basis. In case of any incompliances, the engineer can readily react and investigate, determine, and rectify the cause; rendering the process (of deaeration) effective again.

While the above may sound like a very routine integrity management case (for a seawater injection system), past observations have illustrated that some seawater injection systems fail simply because of the lack of the above performance monitoring, or inadequate effectiveness assessment, or both.

Example Three: For many assets, corrosion under insulation (CUI) is a major integrity threat; combating CUI would require three things:

1. An up-to-date CUI register containing all insulated lines and equipment
2. A practical and efficient CUI strategy detailing how the CUI threat should be mitigated
3. Implementing the strategy using the register

In this case, performance monitoring can be achieved by counting previous CUI failures per annum. The acceptance threshold would be having no annual leaks and/or failures due to CUI. Therefore, the effectiveness of the CUI campaign can be assessed via comparing the previous annual CUI leaks with the accepted threshold or target (of no CUI annual leaks).

3.13 The Corrosion Management Implementation Process

Once both the integrity management measure and the corrosion management concepts have been fully comprehended and personnel have become completely conversant with the corrosion management process, then it is time for implementation. Any corrosion management implementation process should have the following three phases.

3.13.1 The Pre-Implementation Phase

During this phase, the following parameters have to be discussed, clarified, and mutually agreed upon between the two parties (e.g., the company receiving corrosion management consultancy services and the corrosion management consulting contractor or provider):

- The exact scope of work (e.g., the type of systems and equipment included and excluded, well-defined starting and finishing points for the project's boundary or scope).
- The type and extent of input data (e.g., inspection data, design, and process data, available piping and instrumentation diagrams (P&IDs), material selection diagrams (MSDs), and the chemical injection data). It is also of significant importance for the company to advise the corrosion management contractor of the degree of reliability of the various inputs to be provided.
- The company representative(s) who would provide the agreed upon- and required input, and who would liaise with the corrosion management contractor during this phase and beyond.
- The corrosion management contractor representative(s) who would take part in this phase and beyond.
- The type and content of output (e.g., presentation slides, a written report or simply a list of findings) and the extent of analysis and data trending to be carried out on the provided input.
- The estimated duration of the work and the associated costs.

However, if such corrosion management work is going to be carried out within a single company, with its different departments taking part in it (in other words, it is an internal corrosion management exercise or project), then some of the above items may not apply anymore.

The following sections explain the other two crucial components of the pre-implementation phase.

3.13.1.1 Senior Management Awareness Session

It is extremely important that any corrosion management implementation project is preceded by a short (corrosion management) awareness session or presentation for the pertinent senior management. It is the senior management who always approve and allocate resources (i.e., personnel, budget, and time). Therefore, they should be fully aware of the corrosion management concept, process, implementation, benefits, and (the associated project) requirements.

Furthermore, a better-informed or more-aware senior management can mean enhanced, steady, and continuous management support throughout the implementation phase and possibly beyond (i.e., during the post-implementation phase). Conversely, a poorly informed management can downsize, (erroneously) revise, or terminate the corrosion management work or implementation at any time. The author has witnessed a few cases where the management had shelved the completed work–without heeding the findings and implementing the pertaining (corrosion management) recommendations–because of a nonexistent or inadequate initial corrosion management awareness session or presentation.

3.13.1.2 Site Personnel Training

Once the corrosion management awareness session for the management is over, then it is time to provide thorough and comprehensive corrosion management training for the (site) personnel. The selection of personnel for such training should be based on the following criterion: all individuals taking part (either directly or indirectly) in some or all of the three phases of the implementation process.

Table 3.1 lists all the main topics or subjects that should be covered (as a minimum) during such a corrosion management training course. Based on the author's personal experience, the recommended duration of the course should be at least four days and should cover in detail all the topics listed in the table.

Table 3.1. Topics to be covered during typical corrosion management training courses, along with the associated importance.

NO.	SUBJECT	IMPORTANCE
1	Introduction to Corrosion in the Industry	Medium
2	Brief Introduction to Corrosion Engineering Concept	High
3	Corrosion Management Concept Definition	High
4	Distinguishing Between CE and CM Concepts	High
5	Classical Integrity Management View	Medium
6	Corrosion Management Process and Products	High
7	Benefits of Corrosion Management Applications	High
8	Shortcomings Associated with CM Applications	Medium
9	Corrosion Management Implementation	High
10	Asset Corrosion Management Strategy Document	Medium
11	Corrosion Control Matrices	High
12	Corrosion Key Performance Indicators	High
13	Data Management and Communication	High
14	Corrosion Failure Preemption	High
15	Corrosion Cost Optimization	High
16	Leak Register and Failure Investigations	High

Ideally, all training course attendees should also take part in an (open book) assessment at the end of the course to ensure they have become fully and adequately conversant with the corrosion management concept, process, applications, benefits, etc.

To facilitate the preparation and organization of the training course, the listed subjects are associated with a priority index (Table 3.1) to aid the training course instructor or provider to know where the main focus and emphasis of the corrosion management training course should be.

Failure to carry out this type of corrosion management training at the Pre-Implementation Phase, or providing poor and inadequate training would definitely have adverse and detrimental consequences on all three phases of the corrosion management implementation process and their desired outcomes.

3.13.2 The Implementation Phase

During this phase the corrosion management process is implemented across both CE-based and non-CE-based integrity management measures. The main output of this phase should be a comprehensive report detailing all the findings and the associated conclusions and the pertinent recommendations. Ideally for each finding or shortcoming identified, there should be a recommendation, detailing how to rectify or improve that particular shortcoming.

Furthermore, a corrosion key performance indicator table should also be incorporated in the report (where applicable) to facilitate monitoring the performance of the Corrosion Management System or Strategy in place and being improved.

In general, some of the main outputs of this phase are the following items:

- A thorough gap study analysis indicating the missing and deficient components of the pertinent AIMS
- The relevant corrosion control matrices
- The relevant corrosion key performance indicators
- The asset corrosion management strategy document (also referred to as the asset corrosion control document), detailing all the relevant asset details on a system by system basis
- All the associated piping and equipment corrosion loops
- The failure risk assessment results associated with all the identified piping and equipment corrosion loops
- A risk-based inspection scope
- A complete team structure along with the associated job descriptions, minimum competency requirements and reporting lines
- Well-defined reporting requirements (e.g., reporting frequency, content, report producer, report recipients and report template)

3.13.3 The Post-Implementation Phase

As had been mentioned earlier, the corrosion management process is composed of three smaller processes where regular performance monitoring and effectiveness assessment were the last two. These two processes (i.e., performance monitoring and effectiveness assessment) become the main tasks or activities throughout the post-implementation phase, along with periodic or regular communication (i.e., reporting) to ensure that critical asset data (e.g., inspections, chemical treatment, leaks, CP, repairs, corrosion key performance indicators [KPIs], deterioration rates, and remaining life estimates) are disseminated among all pertinent personnel and managers at all times. It is throughout this phase that all the implemented integrity management measures are constantly and continuously being monitored and their effectiveness assessed.

If the assessed magnitude, quality, and effectiveness is not acceptable, then the associated shortcomings, mistakes, or causes have to be immediately identified so appropriate rectifying measures can be planned and carried out.

Post-implementation is also the stage during which recommendations made during the integrity review process have to be implemented. Some recommendations can be immediately implemented. That is, they are implemented within three months after the submission of the final (corrosion management) report. On the other hand, some recommendations can take much longer than three months to be fully implemented.

The recommendations implemented *within* three months are often associated with the non-CE–based integrity management measures, and the recommendations implemented *after* three months are associated with the CE–based ones. That is, the implementation of the non-CE–based integrity management measures is often much easier and quicker (and also markedly less costly) than the implementation of the CE–based recommendations. These are considered as some of the most important advantages of the non-CE–based integrity management measures as compared to the CE–based ones, both throughout the post-implementation phase and in general.

Simultaneously, the worst thing that can happen during the post-implementation phase (and it does unfortunately occur frequently) is the cessation of communication between the asset (personnel) and the individual or the team who carried out the integrity review and facilitated the corrosion management implementation phase.

Such an occurrence, in most observed cases, reduces the whole corrosion management implementation process to an incomplete or truncated exercise with no or minimum outcomes. Furthermore, any produced reports (during the implementation process) will usually end up on the shelves, until the next implementation project, in a few years' time.

Therefore, it is crucial that throughout the post-implementation phase the asset and the individual or the team, who carried out the integrity review and facilitated the implementation phase stay in touch with each other. That is, the company *receiving* the CM services gives the individual or team *providing* the CM services with regular reports and corrosion key performance indicator figures.

3.14 Conclusions

The author's personal experience—gained through dealing with various United Kingdom and international oil and gas integrity management projects—illustrated that what actually distinguished successful and efficient AIMSs from the unsuccessful and inefficient ones were two items.

1. Carrying out the following three processes on a regular basis:
   a. The integrity review process
   b. The performance monitoring process
   c. The effectiveness assessing process

2. Equally weighting the non-CE-based integrity management measures as much as the CE-based ones; therefore, the above three processes would be carried out on both types of integrity management measures.

Therefore, the author incorporated the above three processes and created a new and bigger process and called it the "corrosion management process." This three-stage process along with the flowchart (Figure 3.2), which depicts various CE-based and non-CE-based integrity management measures, comprise the MCMM.

"Corrosion management process" is a rather new term and as such there still exists some confusion and misunderstanding as to what exactly a corrosion management process is. To some individuals, corrosion management is still just a synonym for corrosion engineering while for others it means a risk-based inspection scope. For some other individuals corrosion management is a systematic approach to corrosion and integrity cost reduction. Therefore, it appears that there is still a long way to reach a consensus on the true meaning of the corrosion management concept.

The corrosion management implementation process is composed of three stages:

1. Pre-implementation stage
2. Implementation stage
3. Post-implementation stage

During the pre-implementation stage, corrosion management training sessions for personnel, corrosion management awareness sessions for senior management, and data collection are all critical components. During the implementation phase, data and information reviews, carrying out a gap analysis, forming corrosion loops and carrying out the failure risk assessment process, determining the corrosion control matrices and corrosion key performance indicators are the crucial components. The most important part of the post-implementation process is the implementation of the recommendations made at the end of the implementation stage.

The 12 Management Requirements

4.1 Background

The non-CE-based integrity management measures play a crucial role in both maintaining and enhancing an asset integrity management system (AIMS) in general, and a corrosion management system in particular. This explains why they have been assigned an equal position, or rank, as the CE-based integrity management measures in the MCMM (Fig. 4.1).

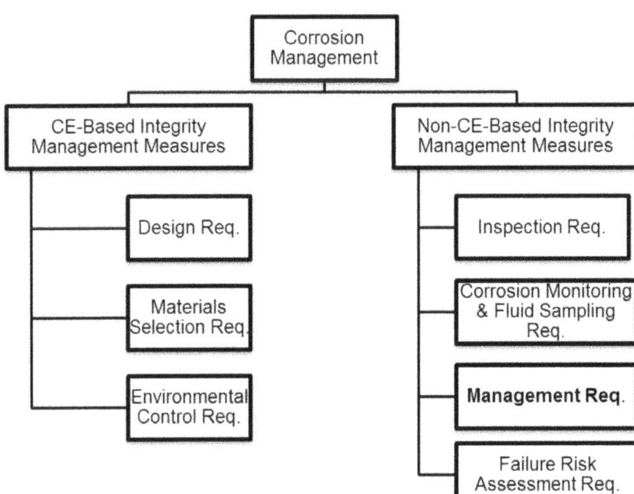

Figure 4.1 The equal position, or rank, of both the CE-based and the non-CE-based integrity management measures in the MCMM.

According to the MCMM, the non-CE–based integrity management measures are divided into four categories:

1. Inspection requirements
2. Corrosion rate monitoring and fluid sampling requirements
3. 12 management requirements
4. Failure risk assessment (or risk-based inspection [RBI]) requirements

Based on the author's oil and gas experience the (12) management requirements have proven to be one of the most important non-CE–based integrity management measure categories. This is because of the fact that in the majority of the corrosion failure or leak cases encountered or reviewed, initially in the U.K.'s offshore oil and gas industry and then elsewhere, these 12 measures had been identified as the main, or at least as one of the main, culprits or root causes.

The same pattern gradually emerged later as the author started providing integrity and corrosion management services for the non-hydrocarbon industries. That is, the same category of integrity management measures was behind most of the observed or reviewed integrity and corrosion issues in such industries. Because of the extremely crucial role the 12 management requirements (Fig. 4.2) play in relation to any AIMS, the whole of Chapter 4 has been dedicated to them.

Figure 4.2 The 12 management requirements.

4.2 Registers

Registers can be defined as documents that list similar items of equipment or those that share the same condition or attribute. Following are some examples of different types of registers:

- Line Register: A list containing all piping within an asset along with their associated numbers, material type, insulation condition, etc.
- Pressure Vessel Register: A list containing the entire asset's pressure vessels and their tag numbers, construction material, lining or cladding, etc.
- Corrosion Under Insulation (CUI) Register: A portion of the line register that only lists the insulated piping and equipment items.
- Deadleg Register: A portion of the line register that lists all piping and equipment deadlegs along with some other useful information such as their type (i.e., design or operation dead leg), orientation, specific inspection (or sampling) requirements, etc.
- Leak Register: A list of all previous leaks containing vital information such as the material that failed, leak location, leak date, last inspection type and date, vicinity to welds, etc. An up-to-date and comprehensive leak register is a valuable input for any failure risk assessment (or RBI) process.

In most of the above cases, a register is often used when an inspection scope, or a specific strategy or a particular chemical treatment has to be devised or planned and then implemented. Accordingly, if the register to be used is nonexistent, not up-to-date, or inaccurate, then items that are absent from the register (or inadequately or erroneously presented) may miss the upcoming–and much needed–inspection or treatment. This would lead to higher corrosion rates, and even leaks and failures over time, within those items missing from the register.

In many AIMSs a piping register or a deadleg register is used as input data for the inspection software to facilitate creating the annual or an ad hoc inspection scope. An incomplete or out-of-date register might cause some high-risk items (which are missing from the register) to never be inspected until they fail or leak. Therefore, it is critical that registers are regularly maintained and updated by designated personnel.

4.3 Strategies

Many assets, because of their age, past integrity management regimes, or current integrity requirements need special attention or treatment for some of their equipment or particular areas. Such a special treatment (e.g., inspection, monitoring or chemical treatment) can be specified and detailed as a separate document, often referred to as a "strategy document."

The following are some examples of such strategy documents:

- CUI Strategy Document: This document details typical CUI issues and threats and how best to tackle them. It can detail an independent or bespoke inspection or monitoring campaign for the insulated items and (if needed) how to prioritize their treatment.
- Deadleg Strategy Document: This strategy focuses on an asset's deadlegs, their prioritization, ranking, and their inspection details or any other details pertaining to corrosion mitigation within such deadlegs.

- Microbially Induced Corrosion (MIC) Strategy Document: MIC (bacterial corrosion) can appear as a complicated integrity issue for some assets because of a multitude of reasons including erroneous chemical treatments, wrong sampling, and lack of competency. However, a clear and comprehensive set of instructions in the form of a strategy can help to reduce MIC and bacterial issues, and improve the integrity condition of the asset.

In general, in some mature (older) assets there might be some chronic and seemingly baffling corrosion or integrity problems that have been going on for years. In such cases, the asset corrosion engineer should identify, clarify, and (as much as possible) quantify these issues during the integrity review process and then if deemed suitable, create a strategy to enhance the mitigation–or even the termination–of such corrosion or integrity problems.

In the absence of such strategy documents and the required strategy itself, these areas would continue to suffer from higher deterioration rates or higher numbers of acute integrity issues. Hence, a strategy or policy that is clear, up-to-date, and bespoke can significantly mitigate a specific type of deterioration within a particular system or area over time, once strategy or policy has been created and implemented.

4.4 Procedures

A procedure (document) can significantly clarify an (integrity-related) activity or process and facilitate its correct and effective implementation. Such implementation can lead to lower corrosion rates, fewer corrosion issues, and an improved chemical treatment. Thus, such a procedure document can positively contribute to the overall AIMS. Simultaneously, the lack of such procedures can lead to more severe corrosion problems over time. The following are just two examples.

1. The Pipelines Hydrotesting Procedure: Pipelines are hydrotested prior to commissioning, and an erroneous or even a nonexisting pipeline hydrotest procedure can inflict significant integrity damage on a pipeline before it enters the operations phase. Therefore, it is important that a procedure is produced for pipeline hydrotesting, clearly explaining the hydrotest fluid requirements in terms of composition, oxygen content, solid content, pH, chloride content, maximum hydrotest duration, etc.
2. The Wall Thickness Anomaly Procedure: Wall thickness anomalies are areas that suffer from higher corrosion rates and lower remaining lives. Thus, they pose a greater integrity threat and as such should be treated with more care and caution–that is, they have to be inspected more regularly and in more detail. Also, if their wall thickness falls below a certain thickness threshold, a repair or replacement has to be planned for them *before* these areas either leak or fail. Such a treatment should be written in detail and in very clear language to avoid any possible confusion in inspecting, identifying, reporting, and finally rectifying wall thickness anomalies.

Every effort has to be made to ensure procedures are fully adhered to and implemented. Simultaneously, during the integrity review process, the asset corrosion engineer has to ensure that the required procedures already exist and that the procedures remain clear, relevant, up-to-date, practical, and effective at all times.

4.5 Data Types and Processing (Data Management)

To carry out successful asset integrity management in general and asset corrosion management in particular, one would rely on and use various data types as their input. Such inputs are then processed (i.e., analyzed or converted) to produce the desired or intended outputs. The inputs and outputs are then used to create an integrity image or picture of the asset where everything can be clearly observed, and in as much detail as possible. The term "everything" refers to both CE–based and non-CE–based integrity management measures that are either functioning as expected, or are nonexistent, defective, or not functioning per the specified expectations or design.

One of most prevalent issues observed regarding numerous assets and their AIMSs is often limiting the (input) data types to inspection data only–in particular when carrying out the integrity review process or the failure risk assessment (or RBI) process. Such an erroneous approach would compel the asset corrosion engineer to view the existing integrity picture or the incumbent integrity issues from a rather tight angle. In other words, in such cases, the engineer is only able to view a relatively small portion of the overall integrity picture associated with the asset.

For any oil and gas asset type, there is a multitude of data types and each type acts as a small piece of a puzzle that will only show the whole and complete (integrity) picture when all its pieces have been put together. Thus, the more data types used, the more complete the puzzle looks and the better one can observe, determine, or identify the existing integrity issues. Consequently, better rectification or integrity solutions can be devised, planned, and then implemented.

Conversely, there are many oil and gas assets where their owner or operators have been trying for years to rectify the existing corrosion problems by implementing some integrity solutions. However, because the full integrity picture (i.e., the main root cause or the principal deterioration mechanism) had not been created or completed in the first place, then the implemented activities were either not relevant, or comprehensive, or effective, or adequate. Therefore, it is of utmost importance that during the integrity review process the asset corrosion engineer determines the following (input) data prerequisites:

- Type of required data (e.g., design, operation, process, and inspection)
- Source or discipline to provide the data
- Reporting frequency
- Reporting format
- Data reliability
- Data processing requirements
- Dissemination requirements (e.g., frequency, format, and recipients)

Any existing shortcomings with the data types, their extent, reliability, and processing can have adverse repercussions on the (asset integrity) picture that can be created. Consequently, the less complete and clear, or the more blurred and vague the picture is, the harder it becomes to clearly observe, identify, quantify, and finally rectify the incumbent (integrity) issues, problems, and shortcomings.

Therefore, data types and their (proper) processing have a significant and crucial influence on how one carries out asset integrity management. Figure 4.3 illustrates the overall data management process.

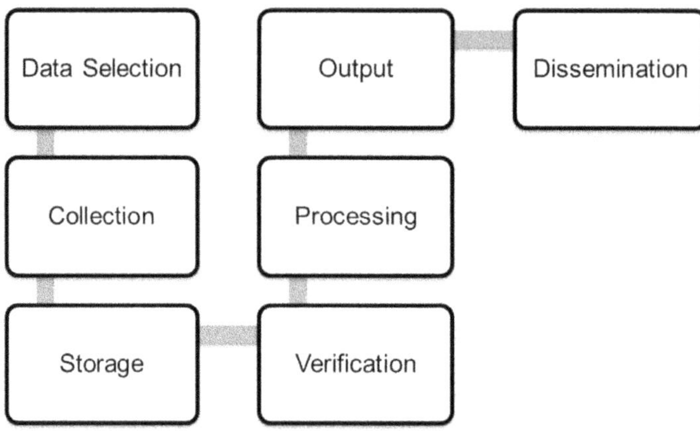

Figure 4.3 The overall data processing (or management) process.

Last but not least, to create a better picture of the integrity condition of an asset and its various equipment, as many data types as possible should be used. Figure 4.4 illustrates such data types.

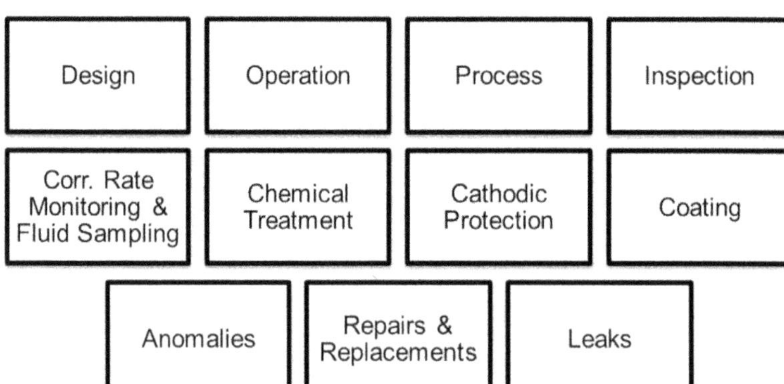

Figure 4.4 Recommended data types for enhanced integrity review and failure risk assessment (or RBI) processes.

The ultimate or ideal objective of data management should be making all the pertinent and required (input) data digital. Such data digitization would facilitate future data trending, analysis, and the eventual decision-making process when artificial intelligence takes over such activities. Some oil and gas operators are already contemplating such a move.

4.6 Databases

Having access to various useful and up-to-date data and information, which are kept in an orderly format, can help the corrosion engineer to make better engineering judgments and technical decisions. Simultaneously, the lack of some vital or up-to-date data can adversely affect engineering or technical assessments and exacerbate an AIMS's performance. Thus, depending on the nature, conditions, and requirements of an asset, creating and maintaining up-to-date databases can be extremely useful. One of the most important and desirable attributes of such databases is the ability to filter or instantly process data with ease. Such a feature, for example, enables an asset corrosion engineer to more conveniently identify the existing deterioration trends. The following are two examples of such databases.

- Pipe Specification Database: Such a database would include all the relevant information for all the pipe specifications used throughout the asset. This information would greatly help the corrosion engineer; in particular, during wall thickness calculations, anomaly management, and planning pipe repairs and replacements.
- Anomaly Database: An anomaly can be regarded as a feature or part of equipment that has suffered a higher deterioration rate, and its risk of failure (due to various threats) is higher than other areas or equipment. As such it should be monitored or inspected more frequently and more detail would be needed to carry out a more accurate fitness assessment for it. Accordingly, it is imperative that an anomaly database is created, where all pertinent anomaly information and data are stored and kept up-to-date. Lack of such a database can adversely affect a sound technical approach or treatment of the existing anomalies resulting in a higher likelihood of failure.

Note that although there are some overlaps between registers and databases with regard to their data storage, filtering, and processing features, sometimes a register might be referred to as a database or vice versa.

4.7 Documentation

The "documentation" category inevitably overlaps with other categories such as registers, strategies, and procedures; nevertheless, because of its significance, it is listed here as a separate and independent category. For many assets, their integrity management is severely impaired because of the lack of appropriate and necessary documents.

Documentation is a general term and covers a broad range of document types:

- Drawings (e.g., piping and instrumentation diagrams [such as P&IDs] and process flow diagrams [(PFDs])
- Reports (e.g., monthly corrosion reports, failure investigation report, and annual integrity reports)
- International and national codes and standards (e.g., various standards from the American Petroleum Institute and the American Society of Mechanical Engineers)
- Procedures (e.g., fluid sampling procedure and welding procedure)
- Strategies (e.g., deadleg strategy)
- Registers (e.g., piping register)
- Meeting minutes (e.g., monthly corrosion report meeting minutes)
- Technical notes (e.g., material selection upgrade recommendations for a particular system)
- Methodologies (e.g., failure risk assessment [FRA] methodology)

During the integrity review process, the asset corrosion engineer has to ensure that the required or necessary integrity documents already exist and are still fully up-to-date, relevant, practical, and easy to understand and (if applicable) to implement. However, if the integrity review process revealed that a particular document was necessary but missing or nonexistent, then it has to be either found, or compiled or created. The lack of a particular document–or a faulty, out-of-date, irrelevant, or erroneous (integrity) document–can sometimes play a profoundly detrimental role on an asset's integrity and its management.

The author has come across numerous cases where shortcomings associated with various document types had resulted in extensive and often costly integrity damage on the associated assets. In one particular case, prolonged failure to produce meeting minutes for the monthly corrosion meetings (and therefore the ensuing failure to assign rectifying actions to pertinent personnel) had exacerbated the integrity condition of three offshore assets significantly.

Unfortunately, a common mistake among some operators or asset owners is that they inadvertently downplay the role and significance of documentation. This weakens the overall efficiency of the AIMS.

Depending on an asset's size, age, integrity condition, and needs, certain document types can play a crucial role in maintaining and enhancing asset integrity. It should be noted that any possible shortcoming associated with documentation can also have a severe negative effect on the "communication" integrity management measure. As explained later in Section 4.13, the communication integrity management measure is regarded as the most important of the 12 management requirements.

4.8 Asset Corrosion Management Strategy

For many assets, corrosion is a primary integrity threat. To minimize this threat, various inspection, corrosion monitoring, fluid sampling, and chemical injection activities often have to be regularly carried out–along with the design and materials selection integrity management measures, which have already been in place.

The author's experience has illustrated that both the effectiveness and efficiency of such activities can significantly be improved when both adequate and up-to-date data, information, and drawings are available. All such data, information, and drawings can be incorporated together to produce a very useful and practical corrosion-related and asset-based document, which is often referred to as the Asset Corrosion Management Strategy (Asset CMS) document, or the Asset Corrosion Control Document (Asset CCD).

An Asset CMS document is a suite of asset-based information and instructions in the form of various corrosion-related databases, drawings, tables, etc. An Asset CMS document would help the project corrosion engineers to significantly improve their understanding of the asset (i.e., process, operations, design, materials, risk assessment, susceptible areas, etc.) and enable them to make better corrosion engineering and corrosion management judgments and decisions. More specifically this document can help them to both devise and implement enhanced inspection, corrosion monitoring, fluid sampling, and chemical injection activities. Hence, an Asset CMS can be an indispensable tool to the asset corrosion engineers and the asset integrity management team.

In general, an Asset CMS document should be composed of the following three main sections.

- Section One: Contains all system-based data, information, and drawings such as
    - Type of materials or alloys used
    - Fluid composition
    - Existing and expected deterioration mechanisms
    - Existing corrosion mitigation system and measures
    - Applicable inspection methods
    - A brief integrity history or background of the system
    - Number of corrosion loops, their associated risk ranking, and inspection intervals
    - System process flow diagram with superimposed corrosion loops
- Section Two: Contains various integrity-related data and information such as
    - List of all the corrosion rate monitoring systems
    - List of all the various chemicals injected or added
    - List of all locations used for sampling
- Section Three: Contains other pertinent strategies, policies, and any other relevant integrity management tools such as
    - Deadleg strategy

- Asset corrosion control matrices
- Corrosion key performance indicator system

Figure 4.5 is an example of the types of data that can be summarized within Section One of an Asset CMS.

<table>
<tr><th colspan="7">Produced Oil</th><th colspan="2">31</th></tr>
<tr><th>Piping Material</th><th colspan="4">Observed Corrosion Mechanism(s)</th><th colspan="2">Predicted Corrosion Mechanisms</th><th colspan="2">Control Measures</th></tr>
<tr><td>Carbon Steel</td><td colspan="4">CO_2 Corrosion, Erosion-Corrosion</td><td colspan="2">MIC, CUI, Deadleg Corrosion</td><td colspan="2">Corrosion Inhibitor Injection</td></tr>
<tr><th colspan="3">Corrosion Rate Monitoring</th><td colspan="6">Two weight loss coupons at the outlet of export booster pumps</td></tr>
<tr><th colspan="3">Fluid Sampling</th><td colspan="6">One sampling location downstream of LP Separator each train</td></tr>
<tr><th colspan="3">Susceptible Features</th><td colspan="6">Deadlegs, Straights, Reducers, and Welds</td></tr>
<tr><th colspan="4">Recommended Inspection Techniques</th><td colspan="5">Ultrasonic testing (UT) and General Visual Inspection (GVI)</td></tr>
<tr><th colspan="3">Representative Fluid(s)</th><td colspan="6">Oil, Condensate, Gas, and Water</td></tr>
<tr><th colspan="9">Failure Risk Assessment</th></tr>
<tr><th rowspan="2">Corrosion Loop</th><th colspan="4">Threat Level Assessment</th><th colspan="2">Risk Code</th><th colspan="2">Inspection Interval (years)</th></tr>
<tr><th>HAZ</th><th>OPER</th><th>INT</th><th>EXT</th><th>INT</th><th>EXT</th><th>INT</th><th>EXT</th></tr>
<tr><td>LOP1</td><td>H</td><td>M</td><td>H</td><td>L</td><td>1</td><td>4</td><td>4</td><td>7</td></tr>
<tr><td>LOP2</td><td>H</td><td>M</td><td>H</td><td>L</td><td>1</td><td>4</td><td>4</td><td>7</td></tr>
<tr><th colspan="9">Inspection Details</th></tr>
<tr><td colspan="9">Carry out a GVI of the whole system and inspect the relevant anomalies. Carry out UT wall thickness inspection of all the selected features on relevant system piping isometrics.</td></tr>
<tr><th colspan="9">Comments</th></tr>
<tr><td colspan="9">Carbon steel piping immediately downstream of the HP and LP separators has been more susceptible to corrosion than any other piping sections in this system. The piping section downstream of the LP separator in Train 2 has almost used up its corrosion allowance and has to be scheduled for replacement in first quarter of next year. This year's microbial survey revealed that there is bacterial contamination downstream of the LP separator in Train 2.</td></tr>
</table>

Figure 4.5 A sample system data sheet summarizing the required data and information for a typical oil system.

4.9 Corrosion Control Matrices

For almost any oil and gas asset type there are various physical and chemical variables that if they are kept within a predefined range, or below an acceptable threshold, the corrosion or the deterioration rate is minimized. Examples of such physical and chemical variables include

- Physical variables: temperature, pressure, flow rate, potential (i.e., voltage), water cut, and dew point
- Chemical variables: pH, corrosion inhibitor (or any other injected chemical) concentration, and concentration of corrosive constituents

For each variable, at least the following five parameters have to be determined:

1. Activity
2. Responsible person
3. Location (where the activity takes place)
4. Frequency (of the activity)
5. Predefined range or acceptable threshold

Once all the above parameters are determined or defined for the associated physical and chemical variables, then grouping them together will produce a matrix (or table) containing exact information and instructions that are absolutely crucial for maintaining the integrity for one particular system. Putting together all the matrices for all the associated systems creates a set of matrices, which are referred to as the "corrosion control matrices." Such matrices impart further clarity and transparency into any existing AIMS and they are absolutely indispensable.

Ideally, corrosion control matrices should be compiled or created at the asset's design stage, once all the pertinent operational-, process-, and chemical injection parameters have been determined and finalized. Once the asset has been commissioned and is in its operational stage, then the asset corrosion engineer should carry out the first integrity review process to fully review the matrices.

Corrosion control matrices play an absolutely significant and critical role in maintaining asset integrity–provided they are compiled on a timely basis and are properly implemented.

However, in many real cases, the matrices are determined only after significant corrosion and integrity issues, failures, and leaks have occurred often inflicting costly rectifying or repair activities on the operator. Any corrosion management implementation option taken (Chapter 9) should definitely include the creation, maintenance, and updating of corrosion control matrices for the concerned asset. Table 4.1 is an example of a corrosion control matrix compiled for a typical seawater injection system.

Table 4.1 A common corrosion control matrix for activities in a typical seawater injection system (u/s = upstream and d/s = downstream). Bold rows are the selected corrosion key performance indicators covered in Section 4.10.

ACTIVITY NO.	RESPONSIBLE PERSON	ACTIVITY	LOCATION	FREQUENCY	THRESHOLD
1.1	**Offshore Chemist**	**Monitor residual chlorine levels**	**u/s of DA Tower**	**Daily**	**0.5–1.0 ppm**
1.2	Offshore Chemist	Monitor residual chlorine levels[a]	d/s of DA Tower	Daily	0.2–0.3 ppm
1.3	Offshore Chemist	Monitor antifoam injection rate	u/s of DA Tower	Daily	Threshold not practical to measure, Offshore Chemist to monitor tank levels to ensure correct dosing
1.4	**Offshore Chemist**	**Monitor biocide injection rate**	**u/s or d/s of DA Tower**	**Weekly**	**180 ppm**
1.5	Offshore Chemist	Monitor scale inhibitor injection rate	u/s of DA Tower	Daily	No Comment
1.6	**Offshore Chemist**	**Monitor oxygen scavenger injection rate**	**d/s of DA Tower**	**Daily**	**10 ppb**
1.7	Offshore Chemist	Monitor residual oxygen concentration (on-line)	d/s of DA Tower	Six-hourly	10 ppb
1.8	Offshore Chemist	Monitor residual oxygen concentration (CHEMets)	d/s of DA Tower	Six-hourly	10 ppb
1.9	**Offshore Chemist**	**Monitor residual bisulfite concentration**	**d/s of scavenger injection point**	**Daily**	**0.64–1.28 ppm**
1.10	Offshore Chemist	Monitor pH[b]	d/s of DA Tower	Weekly	7.0
1.11	Offshore Chemist	Check oxygen probe membrane	d/s of DA Tower	Daily	N/A
1.12	Offshore Chemist	Monitor residual oxygen concentration	d/s of Injection Pumps	Daily	10 ppb
1.13	Offshore Chemist	Calibrate online oxygen probe	d/s of DA Tower	Weekly	N/A
1.14	Offshore Chemist	Monitor sessile bacteria population[c]	d/s of Injection Pumps	Weekly: Pre- and post-biociding	10^2 per cm^2
1.15	Offshore Chemist	Monitor planktonic bacteria population	d/s of DA Tower	Weekly	1 per cm^3

a. Only where the deaeration is carried out through gas stripping
b. Overdosing of oxygen scavenger and/or use of stripping gas containing CO_2 and H_2S gases can render the treated water pH acidic
c. Preferably using a sidestream located d/s of the DA Tower

4.10 Corrosion Key Performance Indicators

Corrosion key performance indicators (corrosion KPIs) are key corrosion control matrices activities that can be used as indices or indicators to monitor the performance and effectiveness of an AIMS. Regular monitoring of their performance will immediately reveal whether an AIMS has been functioning satisfactorily or not. Corrosion KPIs are also regarded as an indispensable tool in maintaining and improving asset corrosion management.

Normally, any corrosion control matrices document is composed of many different activities that, when carried out on a regular basis, contribute to the overall AIM. However, some of these activities play a more critical role in mitigating corrosion and maintaining the integrity of the asset. The performance of such activities can, in general, better reflect the successes and the shortcomings of the AIMS.

The relevant corrosion KPIs should always be selected from the activities listed in a corrosion control matrices document. Selecting the most appropriate corrosion control matrices activities as corrosion KPIs is a delicate job and is normally at the discretion of the responsible asset corrosion engineer.

Once the appropriate corrosion control matrices activities have been selected as corrosion KPIs, they will usually be calculated on a daily basis (provided that the relevant data are produced on a daily basis) and expressed and reported as "percentage compliance on a monthly basis"; preferably within the asset monthly corrosion report.

To be able to express corrosion KPIs as percentage compliance, each single corrosion control matrices activity selected as a corrosion KPI should preferably have either a defined range or a threshold value. As long as the measured or the calculated value is within the defined range, or has not exceeded the threshold value throughout any particular day, then it is considered as compliant for that day. For a 30-day month, each compliant day is equal to 3.3% compliance. Therefore, at the end of each month, the number of compliant days should be added up and then multiplied by 3.3% to determine the monthly percentage compliance for that particular KPI. Once all the individual monthly compliances are determined for all the selected KPIs, then the average percentage compliance for that month can be determined.

Normally, a compliance target level is also defined and used as a benchmark to evaluate the overall performance of the calculated monthly average compliance. The value of the compliance target level depends on the following parameters:

- Client expectations
- Recent performance of the monthly average compliance
- The asset corrosion engineer judgment

A spreadsheet can be used to perform all the necessary calculations to determine individual KPI levels, their monthly average, and finally to plot the required graphs.

Table 4.2 depicts a typical KPI table in an asset monthly corrosion report where each individual activity, along with its corresponding range or threshold, monthly performance, and the compliance target level are listed.

Table 4.2 A typical corrosion KPI table pertaining to a seawater processing facility.

PERFORMANCE MEASURE	FREQUENCY	TARGET VALUE/RANGE	% COMPLIANCE
Monitor residual chlorine levels	Daily	0.5–1.0 ppm	**85**
Monitor biocide injection rate	Weekly	180 ppm	**90**
Monitor oxygen scavenger injection rate	Daily	10 ppb	**100**
Monitor residual bisulfite concentration	Daily	0.65–1.30 ppm	**65**
Average Compliance			**85**
Compliance Target Level			**80.0**

4.11 Team Structure and Roles and Responsibilities

Corrosion- and integrity management implementation processes are team or group activities. Various individuals and disciplines work together to maintain and enhance asset integrity. To improve efficiency and transparency among team members, there should be a clear team or group structure identifying the head or manager, along with all other positions within the team. Furthermore, to avoid any activity overlaps or oversights, there should be clear and detailed job descriptions for each position listing the pertinent roles and responsibilities.

Potential lacks of such team structure and detailed roles and responsibilities can translate into less cohesive and efficient teamwork, less effective corrosion engineering and corrosion management activities, and hence more corrosion and integrity issues over time.

Listed below is a summary of the five main elements of an integrity team structure.

- Team Structure Chart: There should be a chart showing the structure of the integrity team clearly and transparently. On the chart, all required disciplines and expertise should be highlighted, and company and contractor positions should be clearly distinguished.
- Job Descriptions or Roles and Responsibilities: For any position indicated on the chart, there should be a detailed job description or set of roles and

responsibilities. That is, for any individual position appearing on the structure chart, all the pertinent jobs, tasks, and responsibilities should be listed. Furthermore, the physical boundary for which the individual is responsible should be clearly defined along with the type of systems, equipment, or assets the individual should look after.

- Communication Requirements: Any team structure chart should also be accompanied by detailed communication requirements explaining who reports to whom, on what frequency, how (i.e., meetings, written reports, presentations, etc.), where (i.e., location), other possible attendees or report recipients, report contents, etc.
- Minimum Competency Requirements: Any team structure chart and set of job descriptions should also be accompanied by detailed minimum competency requirements for each individual position, enabling only competent individuals to occupy such positions.
- Corrosion Engineering Position: Some of the visited (by the author) assets where corrosion was a major integrity issue, did not have any corrosion engineers. Therefore, it is imperative that such a position be created within any asset where corrosion is an integrity issue. If the corrosion engineer position has been provided by the contractor, then a set of minimum integrity requirements (such as maximum allowable corrosion rate, maximum acceptable number of process leaks per annum, and minimum acceptable corrosion KPI values) should be stipulated by the company to provide an integrity framework for the contractor corrosion engineer to work, function, and report.

4.12 Competency

Based on author's observations and feedback received from numerous operators and asset owners across the oil and gas industry, the non-CE–based integrity management measure of competency is the second most important Management Requirement after communication. That is, in a significant portion of the encountered or reviewed failure investigations competency—or more precisely the *lack* of competency—had been identified as the principal or one of the main contributing factors or root causes of corrosion or integrity failures.

Thus, it is very important that at the same time the integrity team *structure* is being finalized, all the required *disciplines* are identified. After the disciplines are identified, each individual discipline must be associated with minimum qualification or competency requirements." Enforcing such requirements during recruitment of new personnel or promoting existing personnel would always ensure that only adequately qualified and competent individuals would occupy the vacant or available positions. This approach of enforcing requirements can significantly reduce corrosion and integrity failures that are due to competency issues because as mentioned earlier, a considerable portion of corrosion and integrity failures had been linked to lack of, or inadequate, competency.

The competency integrity management measure has four components:

- Education
- Experience
- Training
- Supervision (or mentoring)

In an ideal situation, all four components are present proportionally. However, under normal circumstances the responsible integrity manager or team leader has to ensure that when components 1–3 are lacking or inadequately present, then at least component 4 is fully and thoroughly present and enforced to compensate for the existing deficiencies in other competency areas until they are improved and have reached an acceptable level.

One very critical point regarding the training component is that some senior integrity managers regard personnel training as a mere cost rather than a well-justified investment in their human capital. In some of the observed or reviewed costly integrity cases and failures, often a timely investment of a few hundred or sometimes a few thousand dollars worth of training could have preempted or prevented a multimillion dollar corrosion or integrity failure.

One of the issues that can further exacerbate a lack of training is training budget cuts due to a multitude of reasons. Senior (integrity) managers should be reminded constantly that better-trained and more competent personnel can make considerably larger integrity cost savings compared to situations when the training budget itself is sacrificed to make cost savings in the overall (integrity) budget or costs.

Timely and proper training–which results in more competent personnel–is the ultimate investment and one of the best ways to make integrity cost savings through enhanced (integrity) decision-making and the ensuing implementation.

4.13 Communication

The lack of, or inadequate, communication has been regarded as the most predominant root cause of corrosion failures and integrity issues across the all three sectors of the oil and gas industry. This predominance is based on the author's personal working experience in various countries, and is also based on numerous corrosion and integrity case studies presented or raised by a significant portion of individuals attending the author's asset corrosion management training courses over the years.

Many costly corrosion and integrity failures have been encountered because of the following:

- Failure to hold periodic corrosion meetings attended by the pertinent disciplines (e.g., corrosion, inspection, pipeline, maintenance, operation and process engineers or representatives)

- Poor and inadequate reporting (e.g., written [hard copy], online, oral)
- Poor, inefficient, and/or impractical reporting formats and templates used (e.g., using very small font- and diagram sizes to save paper consumption)

On the contrary, slight enhancements in the communication integrity management measure can create significant asset integrity improvements—at virtually no extra cost. Enhanced communication among the asset integrity team members, means better-informed individuals who can see and identify the existing corrosion and integrity issues more clearly and transparently. Consequently, they can devise more efficient, up-to-date, practical, and relevant integrity solutions and rectification activities.

Throughout the asset integrity management process, different types of data and information are needed or have to be created, analyzed, and then acted upon. Simultaneously, many lessons are learned and have to be communicated to pertinent personnel or a situation arises where a collective decision-making is required. This communication process also has to occur when a particular discipline has identified a new situation, parameter, or condition with potential repercussions for other disciplines. In all such instances, a clear and effective communication can help to

- Improve or rectify a situation
- Get something done
- Get an important message across or notify somebody of an important event, change, or impending incident

When implementing the corrosion management process, regardless of the implementation option taken (see Chapter 9), the communication integrity management measure should always be included within the AIMS.

In general, the communication integrity management measure is an indispensable requirement throughout the asset lifecycle, spanning stages:

- Design
- Construction
- Transportation
- Installation
- Hydrotesting
- Commissioning and operation
- Decommissioning

However, in this section, the main emphasis is on proper and adequate communication throughout the operation stage. Communication can be in either of the following formats:

- Written (hard copy)
- Slideshow presentations

- Online (email)
- Teleconferences
- Periodic meetings
- A combination of the above

All the above categories can then be further divided into regular and ad hoc types. It is mainly the asset corrosion engineer's responsibility to determine which communication format(s) is needed following the integrity review process. Then their frequency, content, recipients (of the information), etc., can be determined. For example, under the periodic meetings category, the corrosion engineer should decide what integrity-related meetings are necessary, at what frequency, and what their agenda should be. If the potential attendees are located in different locations or regions, then they can decide to establish online (or conference call) communication instead of regular face-to-face meetings, or alternatively they can agree on receiving regular written integrity reports.

The communication requirement is considered as one of the most important and crucial elements or components of the management requirements. It is often one of the many culprits leading into a corrosion issue, failure, or leak. The following are examples of poor communication:

- The operations and process teams or disciplines do not communicate relevant changes (which have detrimental corrosion consequences) to the asset corrosion engineer.
- Inspection findings are not communicated to the operation and maintenance teams.
- Critical corrosion topics, parameters, and ranges are not communicated to the operation or process teams. That is, the operation team does not learn about the critical integrity condition of a piece of equipment or an area of the asset until it fails, resulting in a leak, explosion, contamination, incident, etc.
- Ordered material is different from the purchased material (software, chemicals, and alloys).
- Relevant integrity data are not communicated to the asset corrosion engineer enabling them to analyze the raw data and make crucial conclusions or determine critical conditions.
- Asset corrosion engineers do not communicate integrity findings to other disciplines.

Reporting is one of the main components of any communication system and is composed of the following elements:

- Who reports to whom
- Report content
- Input and output type, accuracy, and reliability
- Reporting frequency
- Report structure and format
- Report preceding the meeting

- Who else (apart from those who attend the regular meetings and receive a copy of the relevant report periodically) receives the report
- Single point of contact for compiling the report

For successful and efficient reporting (and hence communication), all the above elements have to be clearly defined and adhered to.

4.14 Conclusions

The non-CE–based integrity management measures play an extremely critical role in maintaining and enhancing any AIMS. Such integrity management measures are divided into the following four categories:

1. Inspection requirements
2. Corrosion rate monitoring and fluid sampling requirements
3. Management requirements
4. Failure risk assessment requirements

Based on extensive field experience from around the world, the 12 management requirements are an indispensable feature of any AIMS. Thus, the asset corrosion engineer has to ensure that (if applicable) all 12 integrity management measures are present and remain up-to-date, practical, relevant, and effective at all times. Among the 12 management requirements, the communication and competency integrity management measures are regarded as the most crucial ones.

5 Other Important Corrosion Management Requirements

5.1 Background

Chapter 4 was wholly dedicated to 12 non-CE–based integrity management measures better known as the Management Requirements. Based on numerous field experiences, asset integrity management system (AIMS) audits, asset visits, and asset data analysis, these 12 requirements were considered as absolutely crucial and indispensable integrity management measures for any incumbent AIMS. However, there are a few other non-CE–based requirements whose creation or timely implementation would further enhance an existing AIMS:

- A management of change system
- Corrosion management audits
- Failure investigations and learning
- The leak register

This chapter discusses these four requirements.

5.2 The Management of Change Process

Certain types of changes can have a detrimental effect on the incumbent integrity condition of an asset or can exacerbate the associated deterioration rate(s). Such changes vary in nature, but can include

- Design changes
- Operational changes
- Process changes

- Procedural changes
- Mechanical changes
- Organizational changes

Thus, to minimize or mitigate any possible detrimental effects associated with implementing such changes on the incumbent AIMS in general, and the existing deterioration mechanisms in particular, there must exist a procedure or system to ensure that any proposed change is fully studied, analyzed, risk assessed, and only implemented if it has been formally approved according to that procedure or system requirements.

The author has come across various high deterioration rate or failure cases where the root causes were mainly operational or process changes that were not communicated in advance with the asset corrosion engineer or anybody else within the asset integrity management (AIM) team. In such cases, the pertinent operational- or process teams or individuals had made the associated (operational or process) changes assuming that such changes would not adversely affect the asset integrity or have the potential to exacerbate some of the existing deterioration mechanism or rates.

Some of the most predominant changes that are frequently observed, and their ensuing detrimental effects, are briefly mentioned below. Conspicuously, the majority of such changes were communicated to the pertinent asset corrosion engineer only when the first signs of a leak, failure, or the accelerated deterioration rates had been identified.

- Reducing the corrosion inhibitor injection rate to an extent not adequate enough to form a protective film inside the system to mitigate internal corrosion led to accelerated internal deterioration rates.
- Reducing the temperature within a glycol (or any of its derivatives such as triethylene glycol) regeneration unit associated with a gas dehydration system resulted in partial gas dehydration and increased internal corrosion rates downstream of the gas dehydration systems due to the presence of wet gas in such areas.
- Changing service from a mixed hydrocarbon phase to seawater without proper and adequate prior seawater treatment resulted in exacerbation of localized corrosion rates due to the presence of dissolved oxygen and bacteria in untreated or inadequately treated seawater.
- Changing an incumbent fluid sampling procedure (or deviating from an existing one) in such a way that the bottled or sampled fluids are not capped anymore led to (1) aeration of the samples and (2) the killing or deactivation of the bacteria (such as sulfate-reducing bacteria) due to their exposure to air in such a way that their presence can not be detected in the lab where rather traditional bacterial enumeration methods are used. This led to a lab report declaring the system was free of such bacterial types, which resulted in the cessation or reduction of biocide injection leading to increased localized corrosion rates due to bacterial or microbial corrosion.

To further highlight the significance of having a management of change system, it is interesting to know that the American Petroleum Institute Recommended Practice (API RP) 580 [7] attributes the existence of a sound AIMS to three components, one of which is a rigorous management of change system. (The other two components are a comprehensive set of corrosion control matrices [sometimes known as "integrity operating windows"] and a risk-based inspection plan).

Figure 5.1 lists the necessary components or the minimum requirements for any management of change system.

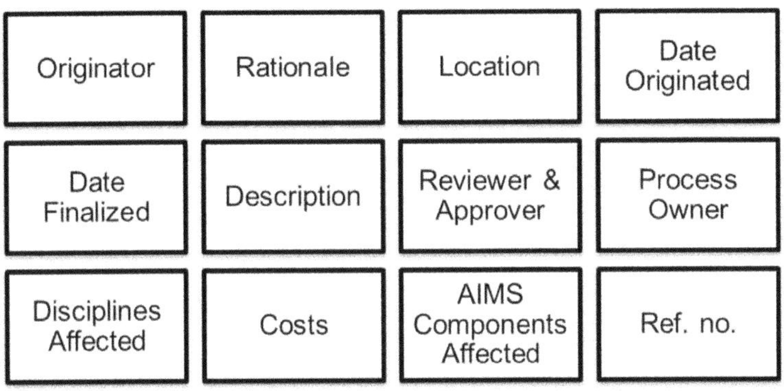

Figure 5.1 Minimum requirements of any particular management of change system.

5.3 The Anomaly Management System

An anomaly can be defined as a physical characteristic of an item under inspection that is considered to be outside a predefined limit or threshold. Therefore, the main two attributes of an anomaly are

- A physical characteristic
- Being outside a predefined limit or threshold

Thus, to be able to define an anomaly, or to be able to determine whether or not an inspected item is an anomaly, one needs to define a certain limit or threshold beyond which a measured physical characteristic becomes anomalous. For example, a wall thickness anomaly can be defined as the minimum allowable wall thickness (MAWT) + 2 mm. Therefore, when carrying out wall thickness inspections, if a measured wall thickness value has either reached the aforementioned defined limit (i.e., MAWT + 2 mm) or exceeded it, then that measured thickness should be regarded as an anomaly. The "defined limit or threshold" is often referred to as the "criteria of non-conformance" or CNC. In other words, the limit set for which a measured thickness value (or any other measured physical characteristic) becomes an anomaly is a CNC. For a wall thickness anomaly

$$CNC = MAWT + 2 \text{ mm} \quad (5.1)$$

Some standards[8] refer to CNC (for wall thickness inspections) as the "Alert thickness" or "flag thickness." Regardless of the terminology used, the CNC value (or threshold) functions as an alarm to highlight the fact that an inspected value is nearing the minimum safety limits–which in the case of the above example is the MAWT value–and thus more frequent inspections or imminent repair or replacement planning might be required.

Examples of physical characteristics include

- Thickness
- Area
- Number
- Angle
- Length
- Weight
- Voltage

Once an anomaly type, or set of anomaly types, have been defined along with their associated CNC values, then an anomaly management system is required for proper and safe management and handling of such anomalies. An anomaly management system can be defined as a set of minimum requirements to be met in the reporting, assessment, and control of anomalies.

The main objective of an anomaly management system is to support an incumbent AIMS by acting as a source of information, guidelines, and procedures on anomaly management. An anomaly management system contributes to maintaining high standards of safety and plant availability throughout the Asset's lifecycle.

It is also the objective of the anomaly management system to detail the minimum requirements to be met in the reporting, assessment, and control of anomalies identified as a result of work undertaken in support of technical integrity. An effective and functioning anomaly management system can significantly improve the failure preemption capability of an operator or organization.

Figure 5.2 shows various anomaly types associated with the aforementioned physical characteristics, which can be used by any inspection department for managing anomalies. However, to be able to manage such anomalies, one has to initially define a CNC value (similar to the way a wall thickness anomaly had been defined earlier as CNC = MAWT + 2 mm) for each anomaly type, which facilitates identifying anomalies in the first place when a particular physical characteristic is being measured.

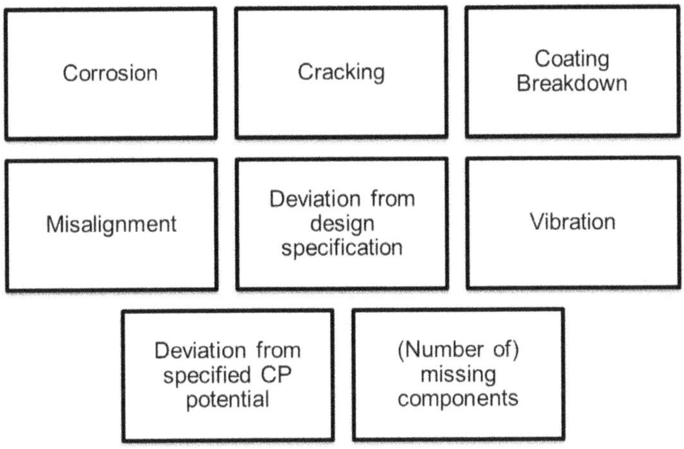

Figure 5.2 Examples of various anomaly types.

5.4 The Corrosion Management Audit Process

The main aim of a corrosion management audit process is to provide an objective and independent assessment and comparison between an incumbent corrosion management system and an ideal one. Once the assessment and comparison are complete, then the findings can enable the identified corrosion management enhancements to be carried out or implemented.

Thus, the main aim of the corrosion management audit process is to determine how effective, efficient, relevant, up-to-date and practical the incumbent corrosion management system is. In other words, the corrosion management audit process can be regarded as a benchmarking exercise or campaign, which is carried out to highlight the existing shortcomings associated with an existing corrosion management system and then to rectify them.

The corrosion management audit can be carried out either internally or externally. In the former case, an individual or a group from the same organization (but preferably from another part or section) carries out the audit, whereas in the latter case, an independent individual or organization carries out the audit. However, it is the author's experience that the objectivity, clarity, and accuracy of *external* corrosion management audits can be far superior to *internal* audits. The main reason for this observation is the leniency often exercised by colleagues when they audit each other, and the tendency to minimize negative reporting within one's own organization. In such cases, various incumbent (corrosion management) shortcomings are not adequately identified, highlighted, and reported.

The corrosion management audit should always follow a structured process or path that can be easily achieved through using a well-thought out and comprehensive corrosion management questionnaire. Such a questionnaire can be divided into the following sections:

- General section (covering commissioning year, remaining design life, requirements for life extension studies, corrosion allowances used, number of past failures, etc.)
- Mitigation section (covering the incumbent and predicted deterioration mechanisms and the associated mitigation to control and minimize such deterioration, the existing deterioration rates, performance of the existing mitigation measures, etc.)
- Organization section (covering the team structure, roles and responsibilities, the associated job descriptions, reporting lines, the required minimum competencies, etc.)
- Failure risk assessment, or risk-based inspection (RBI), section (covering whether the inspections carried out are risk based or not; if they are risk based, then what is the methodology being used; if they are not risk based, then it has to be determined if the inspections are consequence based, condition based, or time based; the risk levels associated with different items of piping and equipment; the allocated inspection intervals with regard to the associated remaining life values; etc.)
- Communication section (covering what types of communications exist, if they are ad hoc or regular, what parameters are reported frequently, if corrosion key performance indicators [KPIs] exist and reported, content of the written reports, disciplines having access to such reports, etc.)
- Documentation section (covering what codes, standards, recommended practices, drawings, procedures, and so on are used, how often they are updated or revised, etc.)

In general, one should remember that any corrosion management audit process should cover both the CE-based and the non-CE-based integrity management measures; in reality, however, most audits only concentrate on the former and ignore or neglect the latter. Figure 5.3 is a further reminder that any corrosion management audit process should cover both integrity management measure types and their respective parameters or variables. For example, when carrying out an audit on the inspection requirements, the associated inspection parameters or variables to be covered in the audit process include

- Inspection techniques used
- Inspection type (risk base, consequence based, or other)
- Inspection frequency
- Inspection extent or area
- Susceptible features
- Data reliability and handling
- Competency of inspection engineers

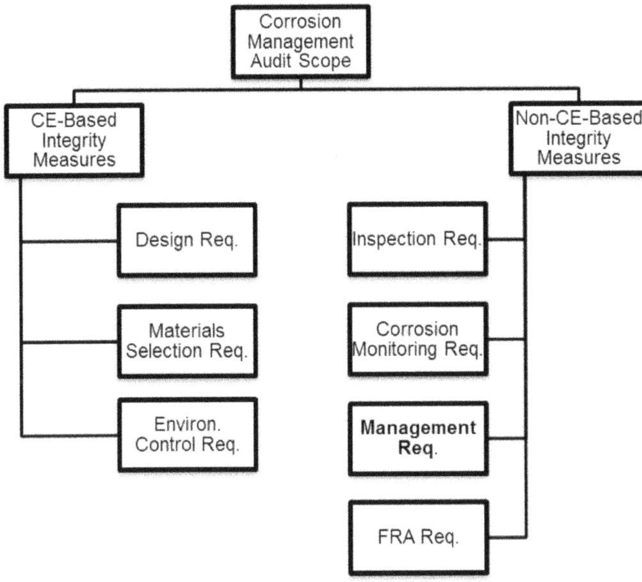

Figure 5.3 The scope of any corrosion management audit process should cover both integrity management measure types.

The depth or audit detail depends on whether the corrosion management audit being carried out is qualitative or quantitative. A qualitative audit is faster, cheaper, and less detailed compared to a quantitative one. Figure 5.4 lists the main attributes of each type.

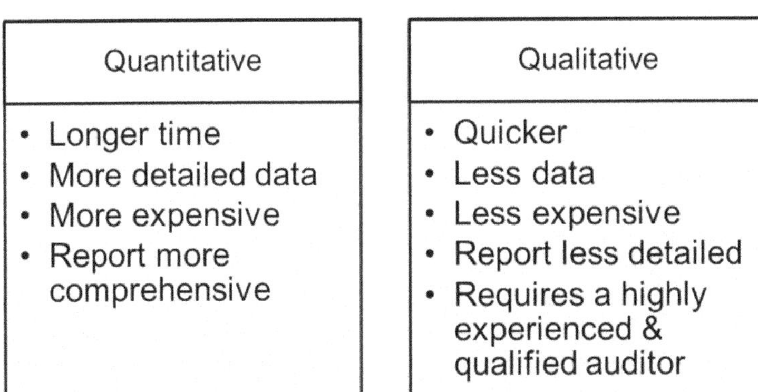

Figure 5.4 Comparison of quantitative and qualitative audit types.

Regarding any corrosion management audit process, there are other important points to consider:

- Any audits carried out should include or cover both the company and contractor positions and activities.
- The individual (or the team) who carries out the audit should be fully conversant with the audit process and the corrosion management concept and system components.
- The audit should follow a planned schedule.
- Identified corrective actions should be registered and clearly highlighted.

5.5 Failure Investigations, Leak Register, and the Learning Process

A failure (i.e., loss of containment) or leak can inflict various costs on any asset owner or operator:

- Wasted product costs
- Production shutdown costs
- Repair and replacement costs
- Insurance and legal costs
- Costs to an asset owner's or operator's reputation (i.e., both tangible and intangible costs)

In spite of such costs, any single failure or production leak has the potential to help personnel identify the root cause(s) behind the associated failure. Therefore, once an asset has suffered from a failure or leak incident, the asset corrosion engineer has to ensure that the main causes of the failure are fully investigated, identified, and rectified so such incidents are not repeated in the future. Thus, behind each (corrosion or integrity) failure is a lesson that should be fully learned and well communicated among the pertinent personnel so the likelihood of reoccurrence of similar leaks is minimized as much as possible through rectification and improvement actions. Such actions can be either CE-based, non-CE-based, or both.

Therefore, it is absolutely indispensable to learn the embedded lessons associated with each corrosion and integrity failure to avoid suffering similar failures over time. Such an approach further highlights the criticality of carrying out failure investigations following each leak and also having a leak register (database) to be able to process and analyze the leak data to better determine both the causes and the required rectification or improvement actions.

5.5.1 The Failure Investigation Process

As previously explained, any process fluid failures or leaks inflict various costs on an operator or asset owner. However, once the failure has been identified, fully analyzed and investigated, and rectified (e.g., through repair or replacement) it can help the operator to investigate the problem in greater detail with particular emphasis on the following areas:

1. More precisely determine the incumbent deterioration rates
2. More accurately identify the associated deterioration mechanism(s)
3. More clearly highlight the existing shortcomings associated with the
    a. CE-based integrity management measures (e.g., a deficient chemical treatment, cathodic protection (CP) system or coating)
    b. Non-CE-based integrity management measures–in particular, the shortcomings pertaining to the (alleged) risk-based inspection regime; the shortcomings might not have flagged the failed location prior to its developing into a leak during the last inspection carried out on that particular location, item, etc.
4. Learn more about the possible failure modes (pinholes, ruptures, or cracks)
5. Improve understanding of what is going on inside the piping, pressure vessels, tanks, etc.; this would be in cases when the failed item has to be removed and replaced, or when the failure provided the operator with an opportunity to enter or observe the inside of a piece of equipment more meticulously.

The above five items are actually potential *benefits* associated with a failure, so it stands to reason that failure investigations after (all-, certain-, or major) failures should be carried out. When carrying out failure investigations it is necessary to take into consideration both the CE-based and non-CE-based integrity management measures and root causes. Often, a costly failure could have been due to a simple non-CE-based integrity management measure such as inadequate competency, inadequate communication, or inadequate corrosion KPIs.

It is imperative that carrying out failure investigations and compiling the associated report becomes mandatory within operating companies. Failure to do so can result in recurring failures with increasing extent, severity, and deterioration rates–in particular, as the assets age further and approach their end-of-design life.

5.5.2 The Leak Register

Creating and maintaining a (hydrocarbon) leak register can further facilitate failure data analysis and trending. Furthermore, the leak register database can also be used as one of the input sources for the FRA process, where a higher number of leaks (associated with a particular corrosion loop, equipment, or system) can be translated to a higher risk code for that particular loop, equipment, or system.

Another critical aim of having and maintaining a leak register is to use the relevant leak data in a few different ways:

- Identify the associated deterioration mechanism(s)
- Determine existing shortcomings associated with the incumbent (corrosion) mitigation measures
- Identify inspection-related shortcomings that prevented detecting the failure before it occurred

Therefore, for any leak register (database) to be able to perform the aforementioned function it has to have the following sections as a minimum:

- Date (when the leak was detected)
- Material (the type of metal or alloy that leaked)
- Mode of failure (pinhole, crack, or rupture)
- Orientation (or "clock position" of the leak)
- Location (with respect to the immediate upstream equipment or any upstream devices that cause any operational or process changes farther downstream)
- Vicinity to the nearest weld
- Type of equipment suffering from the leak (e.g., piping, pressure vessel, weld, tank)
- Design life versus actual life
- The last inspection date
- Deterioration mechanism(s) contributing to the leak
- Estimated deterioration rate
- Internal or external environment causing the leak
- Existing mitigation methods (which might not have performed well, leading to the leak)

The significance of maintaining a leak register is even greater for older or more mature hydrocarbon assets where failures or corrosion leaks are a rather frequent occurrence. Through proper leak data analysis, processing, and trending, an asset corrosion engineer can identify the incumbent culprits (both CE-based and non-CE-based integrity management measures) behind the leaks and begin eliminating the causes leading to a gradual reduction of both leak numbers and their pertinent costs in the associated mature assets.

5.5.3 The Importance of Learning from Past Incidents and Mistakes

Each failure incident or leak provides the affected operator with a unique learning opportunity and potential improvement or enhancement options associated with certain areas or disciplines (e.g., production, process, inspection, and maintenance teams). Therefore, once the pertaining failure investigation has been carried out and the associated report formally issued, it then has to be disseminated among all the relevant individuals, disciplines, teams or sections that can potentially benefit from it.

Sharing a failure investigation report and the associated findings and lessons learned can be done in various levels:

- Asset level—exclusively within the asset where the failure occurred
- Production company level—within the whole operating company that owns more than one asset (and its other assets can benefit from the findings of their colleagues working on the affected asset)
- Ministerial level—in countries where this is applicable, within the whole Ministry of Petroleum, which often owns numerous assets across the country
- National level—among all various industries existing within the national boundaries
- Regional level—among all neighboring operating or production companies enjoying data sharing agreements
- International level—rendering the findings easily accessible to any individual, group, company, industry, and country that might be interested in such information

One very important point to remember is that once an operator or asset owner has incurred various costs associated with a particular failure, then the best way to help others avoid repeating such failures and incurring similar costs is to selflessly and generously share the lessons learned after each failure.

5.6 Conclusions

There are three non-CE–based integrity management measures plus 15 non-CE–based ones (i.e., inspection requirements, corrosion monitoring and fluid sampling requirements, 12 management requirements, and FRA requirements). Therefore, there are a total of 18 CE-based and non-CE-based integrity management measures. In addition, there are a few corrosion management requirements that further supplement these:

- The management of change process
- The anomaly management system
- Failure investigations
- Leak registers

It is absolutely important to learn the embedded lessons associated with each corrosion and integrity failure to avoid suffering similar failures over time. Thus, failure investigations and leak registers play critical roles in helping operators learn from past mistakes and shortcomings so they can continuously improve their incumbent AIMSs.

An Introduction to Failure Risk Assessment (FRA)

6.1 Background

An asset integrity management system (AIMS) and its subsets offer significantly improved mitigation against corrosion when they are risk-based, in comparison to systems or models that are not risk-based.

The MCMM presented in Chapter 3 is a risk-based one. That is, when carrying out or implementing its various components (both the CE–based and the non-CE–based integrity management measures), the concept of risk is always present and taken into consideration. In other words, corrosion management model components such as design, inspection, corrosion rate monitoring, communication, and competency are devised and implemented in such a way to reduce the associated failure risks as much as possible. Such failure risk reduction can lower the incumbent deterioration rates and gradually minimize the annual number of corrosion failures or leaks. That means, implementing a risk-based corrosion management approach would further improve an operator's corrosion failure preemption capability.

The most important point, however—when embedding the risk concept into any asset integrity management system—is to remember that a risk-based system or program should not only be limited to the CE–based integrity management measures, but it should cover the non-CE–based ones as well. For example, when applying a risk-based approach to the communication integrity management measure, one has to determine or assess how total lack of, or poor, communication—including regular reporting—would adversely affect the incumbent AIMS. Thereafter, the communication and/or regular reporting has to improve to such an extent that any failure risks due to lack of, or poor, communication are either fully mitigated or minimized as much as possible.

One of the areas in which a risk-based approach has significant applications is the inspection integrity management measure. Inspections are mainly carried out to determine the integrity condition of assets and their associated equipment. There are various inspection types that use different criteria to create the pertinent inspection scope. Risk-based inspections (RBIs) are believed to be one of the most efficient types of inspections, based on the author's numerous site visits and asset integrity audits he carried out. Therefore, this entire chapter is dedicated to discussions pertaining to inspection, risk, and risk-based inspection concepts and applications.

6.2 FRA or RBI?

To some users the term "risk-based inspection" refers to a process defined by incorporating *some* of the five processes itemized below–although the level of incorporation varies according to each user's perception of the RBI concept and its scope. However, for many users, operators, or asset owners RBI is viewed as a process composed of *all five* processes:

1. Collect data and information
2. Determine the incumbent deterioration mechanism(s) and the associated
3. deterioration rate(s)
4. Determine the probability of failure and the failure consequence
5. Determine the risk of failure due to corrosion, erosion, and cracking
6. Determine the inspection, monitoring, and mitigation requirements

The "all five processes" view is further supported and promoted by the API RP 580 document,[7] which is structured mainly around the five processes. This document uses the term "risk-based inspection" throughout and refers to the above five processes as the indispensable components or constituents of any RBI process.

The author believes, however, that both of the above notions are erroneous, because the term "risk-based inspection" actually refers to *inspection activities whose scopes are determined using the risk criterion and nothing more.* Instead, the author prefers to use the term "failure risk assessment" to refer to the aggregate process involving the aforementioned five processes. That is, FRA is the process that evaluates the risk of (corrosion) failure associated with various pieces of equipment through implementing the above five processes. Because one of the main outputs of FRA is an RBI scope, the author considers that FRA is an actual "process," one of whose "products" is an RBI scope. Because of the widespread use of the term "risk-based inspection," however, it is considered as a synonym for the FRA process throughout this book; therefore, these two terms are used interchangeably herein, in spite of the fact that the latter is considered the correct term.

6.3 The Inspection Basics

There are many reasons for carrying out inspections of oil and gas assets and the equipment therein. One of the main reasons in some countries is a legal requirement. For example, the Health and Safety at Work Act (passed by the U.K. Parliament in 1974) requires that all of the United Kingdom's companies and employers should provide and maintain a safe and healthy working environment for their employees.

Carrying out inspections is one of the main ways to determine equipment integrity condition and hence, how safe they are to work with or operate. Inspections are also carried out to achieve a number of other goals:

- Determine or estimate the deterioration rate and then the remaining life of the equipment
- Identify equipment faults, defects, and missing parts
- Identify the primary deterioration mechanism(s)
- Identify anomalies (inspected items, where the inspected parameter, e.g., thickness and length, is outside an accepted defined limit)
- Determine equipment functioning and efficiency
- Detect leaks

6.4 The Evolution of Inspection Types

Over time, four different inspection types have evolved (one after the other) with the main aim being to improve inspection efficiency and also to enhance the failure preemption capability (through carrying out inspections). The distinguishing factor among these four types is the criterion used to determine and create the inspection scope. Below are the four inspection types and the associated criteria (used to create the inspection scope or choose the right locations, items, and equipment):

1. Time-Based Inspection: In this type of inspection, certain points or locations are selected, often shortly after an asset's commissioning phase (or even sometimes at the design stage, once all the equipment design details are finalized and sent for construction) and are then inspected at fixed time intervals throughout the operation phase. Ideally, as the asset ages and corrosion-related issues and leaks increase, both the scope (i.e., the selected locations) and the time interval should be reviewed and revised accordingly to enable the inspection team to identify those locations that might become a corrosion leak in the foreseeable future. However, because of inadequate inspection resources, some of the assets visited used a fixed (time-based) inspection scope over the years with minimum scope revisions leading to progressively exacerbating corrosion failure preemption capability; leading to more corrosion failures and leaks.

2. Condition-Based Inspection: In this category, the criterion used for determining both the inspection location and the inspection interval is the condition of the equipment–in other words, the associated deterioration rate (hence the estimated remaining life) of the pertinent equipment. This inspection type is also referred to as the "rule-based" inspection where the inspection interval is a fraction of (often equal to, or less than half of) the remaining life of the associated equipment. To be able to use this inspection type, one has to be able to determine or estimate the incumbent deterioration rates, hence the associated remaining life values. Otherwise, inspection intervals can not be specified or determined.
3. Consequence-Based Inspection: In this inspection category, the consequence of failure is the main criterion for determining the inspection scope. According to this inspection category, the higher service classes (i.e., the more flammable and poisonous fluids) require more extensive inspections at shorter time intervals. In addition, parameters such as population size and density, fluid temperature and pressure, equipment size, and fluid volatility influence an inspection scope that is consequence-based.
4. Risk-Based Inspection: In this category, the concept of risk (i.e., the product of probability and consequence, where both are associated with the phenomenon of failure) is the main criterion for determining the inspection scope. To determine an inspection scope that is risk-based, the probability of equipment failure–which is dependent on parameters such as the incumbent corrosion rates and the number of past corrosion failures and repairs–along with the consequence of their failure–which is dependent on the incumbent process conditions, the liquid's flammability and toxicity, and equipment repair time plus the extent of shutdown–are taken into account. If adequate input data are used to create a fully risk-based inspection scope, this type of inspection significantly enhances the corrosion failure preemption capability of the associated inspection program and considerably improves asset uptime and operability.

There are pertinent API Standards that use one or a combination of the above criteria (i.e., time, condition, consequence, and risk) to determine the inspection scope, demonstrated in Table 6.1.

Table 6.1 Comparison of various API Standards and the criteria they use to determine the inspection scope.

Inspection Type/ Standard	API 510[9]	API 570[8]	API 653[10]	API RP 580[7]
Time-Based	Partly	No	Partly	No
Condition-Based	Partly	Partly	Partly	No
Consequence-Based	Partly (PRD)*	Partly	No	No
Risk-Based	Optional	Optional	Optional	Yes

*PRD: Pressure Relieving Devices

Based on the author's numerous site visits and integrity audits, risk-based inspection has been found to be the most efficient inspection type, particularly in pre-empting corrosion failures, among the aforementioned categories.

6.5 The Risk Basics and the FRA Process

The term "risk" is defined as the product of probability (or likelihood) of an event and the consequence of that event:

$$\text{Risk} = \text{Probability} \times \text{Consequence} \tag{6.1a}$$

or

$$R = P \times C \tag{6.1b}$$

In the context of this book, the "event" is always "failure" (i.e., loss of containment) due to

- Corrosion, or
- Erosion, or
- Cracking, or
- Some combination

Therefore, throughout this book, the term "risk" in general means "the risk of failure," but more accurately it is defined as the product of "probability of failure" and the "consequence of failure."

The consequence component itself is a function of hazard and operability. Hazard can be defined as a physical condition resulting from failure and resulting in injury, death, or environmental degradation. Operability is also a physical condition resulting in loss of product or damage to the equipment. Therefore, a hazard (H) can be regarded as the consequence component affecting personnel and environment, while the operability (O) component of consequence is regarded as the financial losses incurred by the operator or asset owner. Therefore, more precisely risk (R, of failure) can be defined in Equation 6.2.

$$R = P \times (H + O) \tag{6.2}$$

Simultaneously, the probability (P) component (or risk) itself is split into internal and external probabilities, i.e., the probability of failure due to internal corrosive environment and the probability of failure due to external corrosive environment, respectively. Thus, the risk of failure for any item or piece of equipment is a function of the following four variables, which are often regarded as the four threats that determine the risk magnitude:

1. Internal probability of failure or internal corrosion threat (P_{int})
2. External probability of failure or external corrosion threat (P_{ext})
3. Hazard threat (H)
4. Operability threat (O)

Therefore, there are two risk values associated with each single item of equipment: the internal risk (R_{int}) and the external risk (R_{ext}).

$$R_{int} = P_{int} \times (H + O) \tag{6.3a}$$

$$R_{ext} = P_{ext} \times (H + O) \tag{6.3b}$$

There are various parameters that influence the magnitude of each of the above four threats. The nature and extent of these parameters depends on various asset attributes such as age, integrity condition, location, fluid types, corrosivity of the internal and external environments, and so on. Figures 6.1 and 6.2 list some of the parameters that influence the magnitude of the four threats associated with a buried hydrocarbon transfer system.

Internal Probability (P_{int})	External Probability (P_{ext})
• Process fluid corrosivity • Total number of anomalies and leaks (due to internal corrosion) • Deterioration rate • Continuous or intermittent service • Sour conditions • Presence of sand or solid particles	• Soil resistivity • Soil humidity • Soil pH • Performance of existing coating • Performance of existing cathodic protection system • Total number of anomalies and leaks (due to external corrosion) • Third party damages

Figure 6.1 Various parameters influencing the magnitude of P_{int} and P_{ext} for a buried hydrocarbon transfer system.

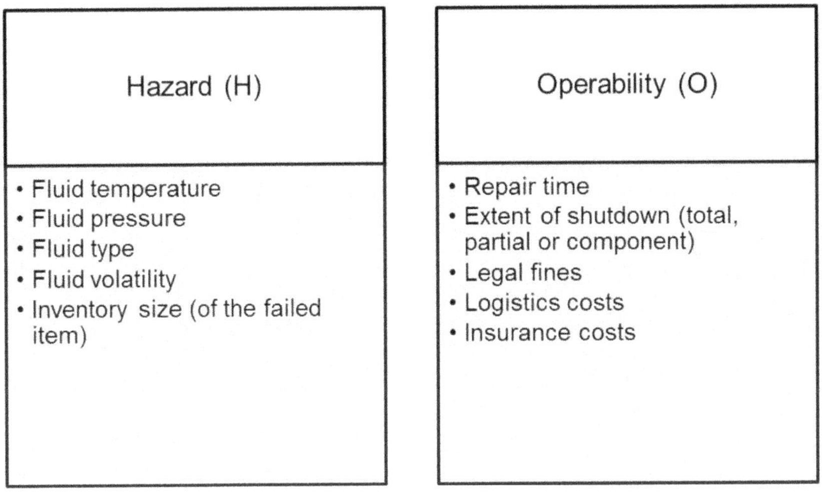

Figure 6.2 Various parameters influencing the magnitude of *H* and *O* for a buried hydrocarbon transfer system.

Therefore, to determine the risk value associated with any asset's equipment, one has to determine the parameters that influence the four threats. The risk value is determined through the FRA process. However, there are three FRA process categories where each category requires different types and amount of data.

1. Qualitative Failure Risk Assessment Process: This category mostly uses engineering judgment and experience to determine the associated risk levels. The FRA methodology is often very simple and straightforward, requiring very little input, and when it is used by someone who is well versed with the asset and its integrity condition, it is a quick and inexpensive option. The output is in qualitative terms such as high, medium, or low. For assets that are moving towards inspections that are risk-based and do not have the financial resources to purchase FRA software, this option can be regarded as a convenient and suitable starting point.
2. Quantitative Failure Risk Assessment Process: This category often uses FRA software and logic models, and requires significantly more input data (compared with the first category) to enable it to carry out the FRA process. Hence, the data entry phase can be rather tedious and time consuming; however, the output possesses more depth and consistency compared with other categories. The most important caveat associated with this category is the creation, marketing, and sale of faulty, dubious, and impractical FRA software, which when used has the potential to inflict more integrity damage (through faulty and erroneous functioning and output) on the asset than good.
3. Semi-Quantitative Failure Risk Assessment Process: In this category, the input data requirement is less than the second category, but more than the first

one. The FRA methodology is often straightforward and can be used by an experienced individual to produce risk results that are as accurate and consistent as those produced by software in the second category. It is also a fast and cost-effective option. This option also offers a great degree of flexibility, associated with its semi-quantitative process, because any new input parameter can be easily and conveniently incorporated into the methodology to improve output accuracy and depth.

The proper implementation of the failure risk assessment process–along with its main product, which is an RBI in scope–can significantly improve an organization's failure preemption capability. This can be translated into progressively lesser numbers of corrosion failures and leaks per annum.

The following items should be expected as the main outcomes or products of the FRA process:

1. Risk values (associated with various corrosion loops and equipment)
2. Inspection technique (e.g., ultrasonic testing, radiography, or visual)
3. Inspection interval
4. Inspection extent or area (i.e., what percentage of an area or what portion of total items are to be inspected)
5. Piping and equipment corrosion loops
6. Process flow diagrams (PFDs)

A very important point regarding implementing the FRA process is the number and types of input data needed to determine the magnitude of the probability of failure. Unfortunately, some organizations mostly prefer to limit their input to only "inspection" type data–even though there are more than 10 input data types that can further improve the accuracy of the calculated (or estimated) magnitude of the "probability" component of the risk.

Thus, both the depth and accuracy of the FRA process further improves. In other words, its failure preemption capability is further improved, leading to progressively less corrosion failures and leaks. Figure 6.3 lists the various input data types that can be used to determine or calculate the probability component of the overall risk.

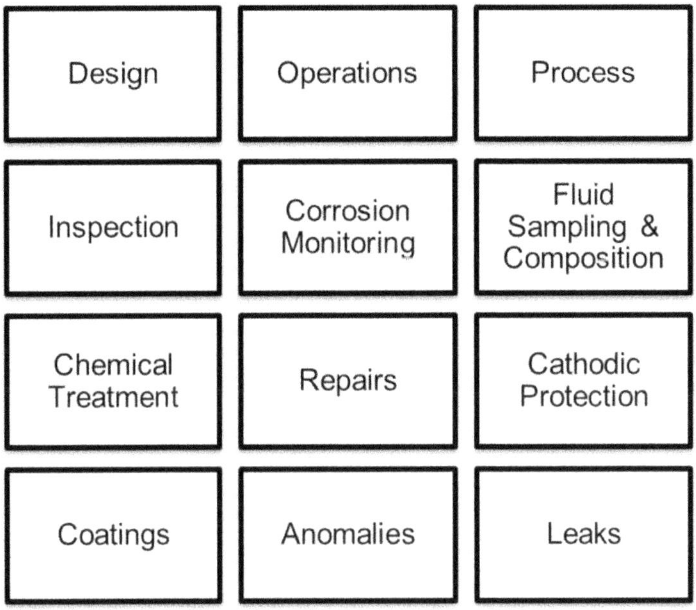

Figure 6.3 Various input data types that can be used to determine or calculate the probability component of the risk value.

6.6 Corrosion Loops

Prior to carrying out the FRA process, the asset has to be "split" into various systems and then each system is split into corrosion loops. Corrosion loops are categorized into two types:

- Piping Corrosion Loops–A collection of piping made from an identical metal or alloy and exposed to exactly the same corrosive environment or conditions. Piping corrosion loops are sometimes referred to as "piping corrosion circuits."
- Equipment Corrosion Loops–Nonpiping items such as pressure vessels, tanks, heat exchangers, etc., can also be treated as equipment corrosion loops. Equipment, based on the various alloy types used to manufacture them and the number of different corrosive environments they are exposed to, can be divided into several corrosion loops. Under this approach, a single piece of equipment is composed of several corrosion loops, where each corrosion loop would be associated with a risk value and the pertinent inspection interval. However, in many cases, a single piece of equipment is regarded as a single corrosion loop with one associated risk value and inspection interval. This second approach further reduces the number of corrosion loops associated

with the nonpiping equipment. The selection of a particular approach for the nonpiping equipment often depends on the resources made available to the inspection teams.

Proper determination of corrosion loops (piping and nonpiping) plays a crucial role in proper inspection planning and then implementing it thereafter. Very often, erroneous boundary allocation to corrosion loops has resulted in too large or too small numbers of corrosion loops where in the first instance (i.e., a large number of corrosion loops), inspection planning was rendered more complicated, and in the second instance (i.e., a small number of corrosion loops) the corrosion failure preemption capability had been compromised. The main reason behind weakening such capability (i.e., the proper determination of corrosion loops) was that the high corrosion rate and low corrosion rate areas were unified; then, the inspection intervals were adjusted based on the low corrosion rate areas–either intentionally or inadvertently.

While determining or identifying corrosion loops is considered an easy and straightforward activity, it is important to carry it out accurately.

6.7 Process Flow Diagrams

Process flow diagrams are an indispensable component of the FRA process or the RBI. Their presence imparts further clarity into any inspection program and its eventual implementation. Below are some of the advantages of these diagrams include.

- A process flow diagram indicates exactly where a system begins and where it ends, i.e., a process flow diagram clearly demonstrates a system's boundaries.
- The diagrams also illustrate all the different equipment types (e.g., piping, pressure vessels, tanks, and heat exchangers) that exist in any system.
- A process flow diagram can also illustrate the different alloy types used to manufacture different parts of a system.
- Once the various corrosion loops have been identified or determined, they can be superimposed on the relevant process flow diagram using various colors.
- Other RBI-related information–such as risk values and inspection intervals–can also be superimposed on the color-coded corrosion loops on the associated process flow diagrams.
- Any other information that might be deemed useful, such as chemical injection points, location of corrosion rate monitoring devises, or fluid sampling points can also be further superimposed on the process flow diagrams.

6.8 Conclusions

An inspection scope can be determined using different criteria such as time, condition, consequence, and risk. An inspection scope that is risk-based has a significantly greater ability to identify high deterioration rate areas or locations, which can develop into leaks in the foreseeable future. Thus, a, RBI scope enhances an organization's failure preemption capability.

The process whose product is a RBI scope is called the FRA process. The FRA process can be carried out qualitatively, quantitatively, and semi-quantitatively, depending on an operator's resources and preferences.

In general the main outputs of an FRA process should include the following items:

- Risk code
- Inspection technique
- Inspection frequency or interval
- Inspection extent or area
- Piping and equipment corrosion loops
- Process flow diagrams

7 Corrosion Management Implementation Benefits

7.1 Background

Proper and timely implementation of the MCMM will bring about numerous benefits. Some benefits can be virtually acquired at no or very little extra costs and resources. This fact by itself should be considered as one of the greatest advantages of the corrosion management implementation.

Implementation benefits can be divided into the following three categories:

1. General benefits
2. CE-based benefits
3. Non-CE-based benefits

This chapter discusses the above benefits categories.

7.2 General Benefits

Corrosion management implementation brings about numerous general benefits:

- Identifying the existing and future deterioration mechanisms
- Enhancing corrosion failure preemption capability
- Optimizing corrosion and integrity costs
- Increasing asset uptime

Sections 7.2.1–7.2.11 discuss general benefits of corrosion management implementation in detail.

7.2.1 Identification of Existing and Future Deterioration Mechanisms

In spite of years of operations and dealing with frequent corrosion leaks and failures, some of the assets (visited by the author) were not certain about the nature and type of the deterioration mechanisms that they were encountering. As such, after many years since their commissioning, they cannot effectively mitigate the associated deterioration mechanisms; this is mainly due to lack of relevant information.

This faulty and deficient (integrity management) approach can be conveniently improved and rectified through corrosion management process implementation. During the integrity review process, asset corrosion engineers are responsible for identifying all incumbent deterioration mechanisms to ascertain whether or not adequate, relevant, effective, and appropriate integrity management measures are already in place. If they are not, then the engineers have to ensure they are created and properly implemented. Simultaneously, deterioration mechanisms, which can be encountered in the foreseeable future, should also be determined and highlighted during the same integrity review sessions. Therefore, planning for their preemption, effective control, and mitigation can begin in advance.

Such an approach minimizes any adverse integrity management surprises by devising the necessary and required integrity management measures in advance; thus, saving time and money in the future through improved knowledge, planning, preparation, and timely implementation (of the associated integrity management measures).

7.2.2 Carrying Out the Integrity Review Process

Unfortunately, many of the oil and gas assets visited did not carry out planned or regular integrity review, and those that did limited the reviews to CE-based integrity management measures. Thus, in many cases, the assets could not identify the main culprit(s) behind poor, inadequate, or inefficient asset integrity management systems (AIMS) when the culprits were associated with the non-CE-based integrity management measures. In some cases, the outcome was increasing severity and increasing numbers of both corrosion leaks and failures.

However, according to the MCMM presented in this book, the first stage in implementing the model is to carry out an integrity review process on both CE-based and non-CE-based integrity management measures. Thus, the incumbent asset integrity management system AIMS is reviewed more thoroughly. Accordingly, required enhancements (both CE-based and non-CE-based) can be identified, highlighted, planned, and implemented; resulting in a more effective and efficient AIMSs.

7.2.3 Improved (Regular) Performance Monitoring

According to the MCMM, carrying out performance monitoring is the second process after the integrity review process. The performance monitoring process enables the operator to have a better and clearer view and understanding about the existing performance of both the pertinent CE-based and the non-CE-based integrity management measures. Thus, any inadequate, poor, or ineffective performance levels can be identified, and then rectified or improved. Such an approach would prevent an unacceptable or inadequate performance level—such as a poor chemical treatment (a CE-based integrity management measure) or an inadequate competency (a non-CE-based integrity management measure)—to remain unidentified or undetected for prolonged periods.

7.2.4 Improved (Regular) Assessing Effectiveness

Once a performance monitoring system is in place and regularly monitors the performance of both CE-based and non-CE-based integrity management measures, the asset corrosion engineer should compare each individual performance value against a predefined accepted threshold or criteria to determine its effectiveness. If the measured performance is found to be unacceptable or inadequate, then the culprit(s) have to be identified and rectified. For CE-based items, a rectifying activity can be better design, improved or upgraded material selection, or enhanced environmental control.

For non-CE-based items, the rectifying activity can be the creation and introduction of those items into the AIMS (if they did not exist beforehand) or, if they already exist, their revision, update, or streamlining. Such an approach will ensure that all crucial integrity management items are continuously monitored and, when required, updated, revised, or improved, resulting in a constantly improving AIMS and fewer integrity-related issues.

7.2.5 Enhanced Corrosion Failure Preemption Capability

Corrosion failure preemption (or failure preemption capability) is considered as one of the most important and indispensable tools and benefits of corrosion management implementation. Through improved determination of probability of failure and deterioration rates, the asset corrosion engineer can establish a corrosion failure preemption capability, or (if it already existed) to markedly improve it.

Through such corrosion failure preemption capability, the asset corrosion engineer can reduce the number of annual corrosion failures and leaks, and possibly achieve the "ideal zero process leaks per annum" target. Reducing annual leak numbers has a direct and significant effect on annual corrosion and integrity costs as well.

In general, corrosion failure preemption can be defined as predicting where, when, and how a failure is going to happen and preventing the failure before it occurs. The following two points have to be taken into account with regard to the above definition.

1. The two main components of the failure preemption process are prediction and prevention (Figure 7.1).
2. Prevention is achieved through one of the following three rectification categories based on the incumbent asset conditions, available resources and mitigation capabilities, and also the nature and severity of the existing defects.
    - Initially through a rectifying action or an enhanced mitigation intended to reduce the incumbent deterioration rates such as
        - Optimizing an existing chemical treatment
        - Improving the control of pertinent operation and process parameters (e.g., flow rates, temperatures, and pressures) per the required design specifications
        - Enhancing the effectiveness of the dehydration process
        - Increasing pipeline pigging frequency (to remove dropped out water in the low points)
        - Improving cathodic protection performance
    - If it is too late for a rectifying action to reduce the existing corrosion rate or to lessen the corrosivity of the environment, then it is time for a repair-type action to prolong the operation life of the affected component until the next planned operation shutdown. Repair methods include
        - Applying clamps
        - Applying wraps
        - Fitting sleeves
    - Finally, if it is too late to plan for and apply a repair-type activity, then it is time for a replacement-type action during an enforced operation shutdown. The shutdown might be only component-based or it can be worse, such as partial or total asset shutdown. However, it would be a controlled shutdown compared to when a failure or leak has been identified forcing the operations to carry out an immediate shutdown.

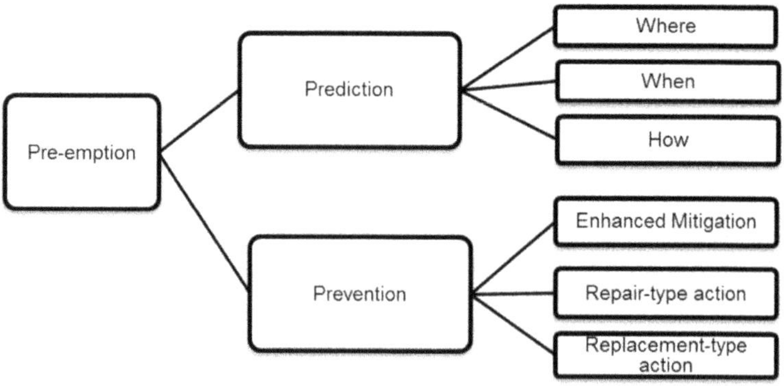

Figure 7.1 Main components of the corrosion failure preemption process.

7.2.5.1 The Prediction Component

To predict the occurrence of a failure means to foresee (with a great degree of confidence), where, when, and how a failure is going to happen. These three parameters are discussed in greater detail:

- **Where (the location):** To predict the location of a likely failure, one has to initially determine whether any deterioration mechanisms can occur at a particular location (e.g., a component or a corrosion loop). Therefore, one has to look into the existing alloy systems and the prevailing internal and external environments to determine whether any deterioration (i.e., corrosion, corrosion cracking, or erosion) is possible or likely at all. Thereafter, areas or locations where any kind of deterioration mechanism is likely should be marked or highlighted. Once all the likely deterioration mechanisms have been highlighted, one has to determine whether any features are more susceptible to any particular deterioration mechanism than other features. For example, if the flowing fluid or product contains sand or any other type of hard solid particle, features such as bends and elbows will be more susceptible to erosion or the erosion-corrosion mechanism than any other feature, such as straights. Some corrosion- or integrity engineers may prefer to carry out the above two tasks during an ongoing failure-risk assessment (FRA) process and not separately. In either case, the areas, locations, and features susceptible to any kind of deterioration mechanism have to be determined and marked.
- **When (the timing):** The timing of a component, system, or corrosion loop failure can be predicted on the basis of the corrosion rate or the remaining life data. The following information can be used individually or together to make this prediction:
  - Ultrasonic testing wall thickness inspection data
  - Corrosion monitoring and fluid sampling data
  - Intelligent pigging data
  - Previous leak data
  - Radiography wall thickness data
- **How (the failure mode):** To determine how a component is going to fail, the asset corrosion engineer has to know the prevailing process conditions and the pertinent deterioration mechanisms (i.e., corrosion, corrosion cracking, or erosion). The next step is to determine whether the indicated deterioration mechanism is going to produce a failure via the following ways:
  - A pinhole (due to pitting or localized corrosion)
  - A rupture (due to uniform corrosion or corrosion cracking or erosion)
  - A crack (due to corrosion cracking)

7.2.5.2 The Failure Prevention Component

Corrosion failure preemption can initially be carried out through a rectifying action, or more accurately through an enhanced mitigation action. However, if for any reason it was too late for such an action to be effective, then the asset corrosion management should resort to first a repair-type action and then ultimately a replacement-type action as detailed in Section 7.2.5 and depicted in Figure 7.1.

7.2.5.3 Critical Factors Affecting Corrosion Failure Preemption Capability

To carry out corrosion failure preemption properly and effectively, several critical factors or parameters have to be taken into consideration. Such factors are explained below.

- **Proper Data Management:** To more accurately predict the time of failure or the failure's location, one has to have access to various reliable data types. Using such data, the asset corrosion engineer can create a clear integrity picture of the asset, system, or corrosion loop. The engineer can also determine or estimate as accurately as possible the incumbent deterioration rate, and hence its remaining life. Such data types include
  - Design
  - Operation
  - Process
  - Inspection
  - Corrosion monitoring and fluid sampling
  - Intelligent pigging
  - Chemical treatment
  - Cathodic protection performance
  - Coating performance
  - Anomalies
  - Repairs
  - Leaks
- **Adequate Communication:** Asset integrity management (AIM) is a team exercise. Many of its components, such as corrosion failure preemption, are carried out more effectively and efficiently when there is adequate communication between the personnel from various disciplines who work on the same team but are probably from different companies or contractors. For example, the corrosion engineer, who is the most likely person to carry out the FRA process and hence, determine or predict the time to failure, has to receive pertinent wall thickness inspection data from the inspection team, as well as corrosion monitoring and fluid sampling data from the operation team. The precise and timely exchange or dissemination of such data markedly improves the quality and accuracy of the output (where, when, and how) produced by an integrity or corrosion engineer, thus facilitating failure preemption.
- **Adequate Competency:** Numerous field observations have shown that many failures occur because of shortcomings in the competency of those in charge of various tasks within the AIM team. A lack or an inadequate mix of education, training, experience, and supervision can lead to erroneous or ineffective rectifying or repair and replacement actions, and therefore to failures. Sometimes, this lack of competency will accelerate the occurrence of failures instead of delaying them.

Any possible lack, or inadequate existence or implementation of such factors will adversely affect both the prediction and prevention components of the corrosion failure preemption process. Such a situation would lead to erroneous indications of the timing, location, and mode of the failure. Furthermore, any rectifying action implemented may not be as effective as intended.

7.2.5.4 Corrosion Management and Corrosion Failure Preemption

The previous sections have, in detail, explained the corrosion failure preemption process, its components, and the critical factors affecting it. A close glance at Figure 3.2 reveals various overlaps and links between this concept and that of corrosion management.

The overlaps and congruity between these two concepts indicate that the timely and proper implementation of asset corrosion management significantly facilitates the process of corrosion failure preemption and its outcomes. In other words, the latter is a product of the former. Once corrosion management is fully implemented across an asset or a system, one should expect to see a marked reduction in (if not the total elimination of) corrosion failures.

These outcomes, products, or benefits are regarded, along with corrosion cost optimization, as the most significant benefits of an asset corrosion management process.

7.2.6 Optimized Corrosion- and Integrity Costs

There are many types of corrosion-related costs and different ways of classifying or categorizing them. In this section, however, the time of failure (during an asset's operation phase) has been used as a chronological reference point for corrosion cost categorization, as illustrated in Figure 7.2. Therefore, under this approach or method, there are two main types of corrosion costs: prefailure and postfailure.

Prefailure corrosion costs are further divided into CE-based and non-CE-based costs, which correspond to CE-based and non-CE-based integrity management measures, respectively. CE-based costs are divided into three smaller subcategories. Some of these CE-based costs (e.g., corrosion allowance and materials selection costs) are closely associated with an asset's design stage. Others (e.g., corrosion inhibitor injection and coating costs) are mostly determined during the design stage but materialized during the asset's operational stage.

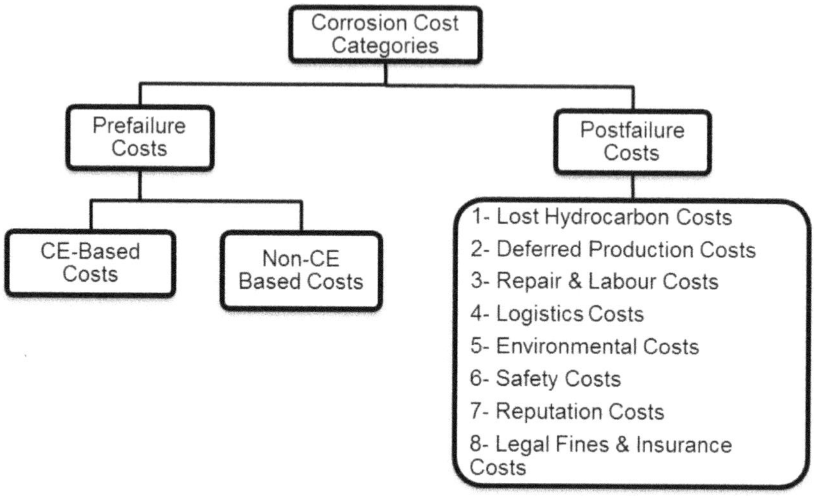

Figure 7.2 Corrosion cost categorization, using time of failure as a chronological reference point for classifying different corrosion cost types.

Non-CE–based costs are divided into four subcategories:

- Inspection costs
- Corrosion monitoring and fluid sampling costs
- Management costs (e.g., producing or updating strategies, procedures, databases, documentation, and communication)
- FRA activity costs

Furthermore, postfailure corrosion costs (Figure 7.2) include various categories:

- Lost product (e.g., hydrocarbon, petrochemicals, and chemicals) and deferred production costs
- Repair and labor costs
- Reputation costs

Once one has fully appreciated different types of corrosion costs, their origins, and the parameters or variables that determine and influence their magnitude, extent, and duration, planning can start on cost optimization. Such corrosion cost optimization has to take place without sacrificing the performance or efficiency of the asset's existing or future CE-based or non-CE-based integrity management measures.

7.2.6.1 Cost Optimization Definition

Corrosion cost optimization can be defined as managing the cost of both CE-based and non-CE-based integrity management measures in such a way that

- Corrosion failures are kept to a minimum (ideally, zero), and
- The efficiency and performance of these measures are not sacrificed, compromised, or adversely affected.

The following points can be highlighted about corrosion cost optimization when defined in this way:

- By preventing or preempting corrosion failures or minimizing the frequency of their occurrences, a significant portion of corrosion costs (Figure 7.2) can be totally avoided. Hence, the overall corrosion cost figure can be reduced markedly.
- Not all corrosion costs pertain to an asset's operation phase; a significant portion is associated with the design phase. A proper design process can therefore play a major positive role in optimizing the overall corrosion costs.
- Because corrosion management by definition incorporates CE-based and non-CE-based integrity management measures, exactly as the corrosion cost optimization definition does, the thorough implementation of corrosion management applications can significantly affect and optimize the overall corrosion costs.

7.2.6.2 Corrosion Management and Corrosion Cost Optimization

A comparison of the corrosion management concept with the corrosion cost optimization concept reveals marked congruity between the two. As mentioned earlier, both concepts incorporate components such as CE-based and non-CE-based integrity management measures. These similarities can mean that adequate and proper corrosion management implementation can influence both CE-based and non-CE-based measures in such a way that not only are potential corrosion failures avoided or preempted as much as possible (i.e., the near total elimination of the postfailure corrosion costs), but also prefailure corrosion costs are optimized—without compromising the extent and effectiveness of such measures.

Eliminating the postfailure corrosion costs (by preempting potential corrosion failures) is achieved through the proper application of CE-based and non-CE-based integrity management measures, which are best optimized by proper corrosion management implementation. The following two sections describe in more detail how such corrosion management implementation can enhance CE-based and non-CE-based integrity management measures. This simultaneously optimizes their pertinent costs, and hence optimizes the overall corrosion costs.

7.2.6.3 Optimizing CE-Based Costs

CE-based costs are divided into the following three subcategories:

- Design (e.g., determination of corrosion allowance)
- Materials selection (e.g., metallic or nonmetallic options)
- Environmental control (e.g., corrosion inhibitor injection)

As illustrated in Figure 2.2, each of these subcategories has a set of variables or parameters that determine the total cost associated with that subcategory and, therefore, the overall CE-based cost.

Determining which of the three CE-based cost subcategories is relevant at any one time is very much dependent on an asset's design stage. Any erroneous results or conclusions regarding the corrosivity of the internal or external environments during the design stage can have significant adverse cost repercussions:

- Thicker or bigger corrosion allowances
- Selection of corrosion-resistant alloys instead of carbon steel
- Inclusion of inner coatings or claddings, instead of relying only on corrosion inhibitor injection for the inside protection of equipment
- Injection of a higher concentration of various chemicals (e.g., corrosion inhibitors)

Opting for such overdesign options because of erroneous fluid sampling or inaccurate compositional analysis of the internal fluid or external environments can have a huge adverse effect on the overall CE-based costs in the design stage. Some components of these overdesign options in the design stage (e.g., overdosed chemical treatment) can also continue well into an asset's operational stage before they are rectified. Furthermore, an underdesigned asset (in terms of corrosion or integrity measures) might appear to have optimized corrosion costs at the design stage, but such assets will suffer from the following corrosion costs after commissioning:

- Increased CE-based costs such as materials or corrosion allowance upgrades, along with higher doses of injected chemicals (to control or reduce an increasing number of corrosion failures due to an underdesigned asset).
- Increased postfailure corrosion costs due to the existence of underdesigned corrosion measures, which engender more corrosion failures to occur with an increasing frequency and also much earlier than expected in an asset's life.

The best ways to optimize CE-based corrosion costs are to first, avoid both overdesigned and underdesigned corrosion (or integrity) measures in the design stage. Second, carry out internal or external fluid sampling and analysis in an accurate and meticulous manner to more precisely determine the corrosivity of the prevailing internal or external environments.

It is crucial that CE–based cost optimization commences during the design stage and continues throughout the operational stage. Therefore, any revisions, alterations, or variations in the existing corrosion measures during the operational stage will directly influence the overall CE–based corrosion costs.

7.2.6.4 Optimizing Non-CE–Based Costs

The non-CE–based corrosion costs are divided into four subcategories, which are associated with the pertaining four non-CE–based integrity management measures. The cost optimization pertaining to each measure is discussed individually in the following subsections.

7.2.6.4.1 Optimizing Inspection-Related Costs

Inspection-related costs are best optimized if the inspection scope is *fully* risk based, whereas a conservative inspection scope would mean extra and unnecessary inspection costs. This is because when a scope is *not* risk based, it may have fewer selected points than a risk-based scope, so it cannot detect high-corrosion rate or high-risk rate areas. There would therefore be a higher likelihood of failure and, thus, higher postfailure corrosion costs.

7.2.6.4.2 Optimizing Corrosion Monitoring and Fluid Sampling Costs

The argument here is exactly as above. A conservative approach to corrosion monitoring and fluid sampling creates unnecessary extra costs. The opposite approach increases the chance of not detecting higher corrosion rates, which can possibly lead to failures and the associated postfailure corrosion costs.

7.2.6.4.3 Optimizing Management Costs

The author's personal experience suggests that many costly corrosion failures are due to the lack of or inadequate existence of items such as registers, databases, communication, and competency. Furthermore, the creation of these items (if they do not already exist) or the updating of existing items can be done, in many cases, at little or no cost. Therefore, producing and updating such items can significantly improve an asset's corrosion management and preempt corrosion failures. As a result, corrosion costs can be significantly optimized.

7.2.6.4.4 Optimizing Failure Risk Assessment Costs

The only cost associated with FRA activities is the cost of carrying them out and their regular update. Therefore, properly planning and implementing the FRA process would optimize such activities and their associated costs.

7.2.6.5 Cost Optimization Misconceptions and Their Repercussions

A significant cost optimization misconception, and therefore a significant corrosion cost problem, is when a CE-based or non-CE-based integrity management measure is reduced, terminated, or postponed without first assessing the possible adverse effects of such measures on the overall corrosion control program and the asset's integrity.

In other words, downsizing or reducing the inspection scope, chemical injection rate, training budget, communication, and so on, without carrying out any prior assessment to determine how such reductions in size, number, rate, or budget can deteriorate the existing integrity situation in the long term. Other similar cases involve delaying the implementation of a particular or crucial integrity management measure (such as repainting a corroded area) without carrying out any risk assessments to assess the potential risks associated with delaying such critical tasks. It is näive to think (or hope) that nothing will happen if the implementation of a much-needed integrity management measure is postponed or delayed. It is also naïve to think that significant cost savings will be made. The author's field experience has shown that in many cases, such approaches will eventually create significantly higher costs due to failures (see postfailure costs in Figure 7.2).

In many observed cases, an instant decision was made to reduce, terminate, or postpone a particular CE-based or non-CE-based integrity measure as a way to make a cost saving. However, such improper and superficial acts often lead to much greater corrosion costs in the long term, with significantly more integrity damage to the asset.

7.2.7 Improved Personnel Safety and Environmental Protection

Through proper corrosion management applications, the integrity of the asset would improve progressively and the likelihood of accidents or failures due to corrosion would decrease. These improvements would enhance onsite safety levels for the personnel, and the asset would offer a greater protection to its surroundings and the environment.

7.2.8 Increased Asset Uptime

Preempting high-corrosion rate features and areas would help in planning the rectifying activities (e.g., painting and chemical treatment) or repair and replacement activities. Thus, there would be fewer and fewer surprise leaks and failures forcing a partial or total unplanned operations shutdown. Consequently, the plant uptime would improve, and features and areas in need of repair or replacement can be taken care of well in advance, during regular or planned shutdowns.

7.2.9 Extended Asset Operational Life Beyond Design Considerations

Timely and proper corrosion management implementation facilitates and promotes improved corrosion control, resulting in enhanced corrosion failure preemption capabilities. Simultaneously, such implementation can further reduce the incumbent deterioration rates to an extent that are below the expected or accepted maximum annual deterioration rates, per the original design specifications.

In other words, through extending the equipment's remaining life, the overall asset operational life can be extended beyond the intended design life.

7.2.10 Optimized Repair, and Replacement Planning and Implementation

Implementing a risk-based integrity management approach after an integrity review process would allow a better allocation of resources and a greater focus on the areas suffering from higher deterioration rates. This improved focus, resource allocation, and increased attention can result in the production of more accurate inspection and monitoring data and lead to more accurate calculation and estimation of corrosion rates and remaining lives. Thereafter, the inspected features with short remaining lives can be assigned a deadline before which they should be repaired or replaced, preferably within the next planned operations shutdown.

Such planning, based on accurate data and calculations, would also remove the need to carry out often unnecessary and costly repairs and replacements in the absence of reliable corrosion rate and remaining life data. With this integrity management approach, one would repair or replace any features or items whose integrity or fitness-for-service condition was suspect, but because of a lack of accurate data, could not be proved otherwise.

Consequently, a risk-based approach after an integrity review process can lead to optimized repair- and replacement activities, and associated costs.

7.2.11 Risk-Based Approach to Asset Integrity Management

The concept of risk (of failure) can be used to better allocate the available resources to the applicable integrity management measures:

- Chemical treatments
- Coating applications
- Inspection campaigns
- Corrosion rate monitoring activities
- Fluid sampling activities
- Competency (i.e., its training component)
- Communication
- Documentation

The risk-based approach to AIM further enhances the efficiency of the incumbent AIMS because resources are allocated to areas, systems, or equipment that need them most. A good example is the significantly enhanced corrosion failure preemption capability achieved through an inspection scope–which is risk-based–as compared with a time-based or consequence-based inspection scope.

7.3 Benefits Associated with CE-Based Integrity Management Measures

The corrosion management implementation process starts at an asset's design stage when determining and specifying the CE-based integrity management measures.

7.3.1 Background

The MCMM is composed of the three processes described earlier in detail. During the design stage of an asset's lifecycle, only the integrity review process can be implemented because there are no acting integrity management measures whose performance and effectiveness can be monitored and assessed, respectively. Therefore, implementing the corrosion management process at an asset's design stage means only carrying out the integrity review process.

At an asset's design stage the CE-based integrity management measures such as design (e.g., determining shape, thickness, and avoiding crevices), materials selection, and controlling environment are determined. In some of the visited operational assets, such integrity management measures had not been reviewed at all following their initial design considerations and prior to their final approval for the next asset lifecycle stage (i.e., construction) causing major and costly corrosion engineering retrofits and rectifications during the asset operation's stage. Therefore, it is important that the design considerations are fully reviewed by the relevant disciplines–in particular, by experienced corrosion engineers to ensure that deficient and faulty corrosion engineering designs are not finalized and approved for construction. A comprehensive corrosion engineering review at the design stage translates to significantly reduced corrosion and integrity costs during the asset's operations stage.

The following three sections discuss the importance of detailed review of the three corrosion engineering components or integrity management measures during an asset's design stage.

7.3.2 Design Review

At an asset's design stage, the pertinent design parameters, such as the following, are reviewed to approve and finalize the best options:

- Shape and geometry of equipment
- Ease of drainage
- No metal mixes
- Crevice avoidance
- Ease of access for repairs and retrofits
- Corrosion allowance thickness

A proper final review of such parameters can ensure lower deterioration rates, hence smaller numbers of process failures and leaks during an asset's operations stage, and accordingly lower repair and maintenance costs.

7.3.3 Materials Selection Review

Similar to the design parameters' review, a thorough review of materials selection parameters listed below has a significant effect on the efficient management of an asset's integrity during its operation stage. Likewise, such a review–along with avoiding under- and overdesigned (materials selection) options–ensures optimized corrosion and integrity costs at both the construction and operation stages by minimizing the required repair and maintenance.

There are four primary materials selection parameters:

- Selection of various metals and alloys
- Selection of various nonmetallics
- Selection of various internal and external coatings
- Selection of various (metallic) claddings

7.3.4 Environmental Control Review

Environmental control (a component of corrosion engineering) can be defined as "corrosion control and mitigation through chemical injections, process parameters variations, and the removal of corrosive agents such as oxygen and hydrogen sulfide." Following are some examples of these controls:

- Corrosion inhibitor injection
- Biocide injection
- Oxygen scavenger injection
- Scale inhibitor injection
- Hydrogen sulfide removal
- Temperature control
- pH control
- Wet gas dehydration
- Wet oil water removal

Like the previous two corrosion engineering subcategories (i.e., design and materials selection reviews), thorough and comprehensive integrity reviewing of the above integrity management measures can result in an effective chemical treatment and process control regime during an asset's operation stage. Such effective chemical treatment leads to optimized corrosion and integrity costs at that stage because of efficient environmental control.

7.4 Benefits Associated with Non-CE-Based Integrity Management Measures

Sections 7.4.1–7.4.5 are dedicated to discussing the non-CE-based integrity management measures.

7.4.1 Background

Non-CE-based integrity management measures are equally important as the CE-based ones. Although unlike CE-based measures, in that they do not directly affect the incumbent deterioration rates, they do positively influence the functioning and overall efficiency of an AIMS. Thus, implemented enhancements associated with this category of integrity measurement measures can indirectly affect the incumbent deterioration rates by reducing them and eventually decrease the number of annual corrosion leaks and failures.

The most important attribute of the non-CE-based integrity management measures is that in many cases their creation, updating, revising, or enhancing requires no or very little resources. Thus, their biggest benefit is that they can be introduced, incorporated, updated, or improved with negligible resources and more importantly—no operation shutdowns. This is the complete opposite of when incumbent CE-based integrity management measures have to be improved, which can be very costly and time consuming, and they require operation shutdowns.

7.4.2 Inspection Requirements

The best type of inspection to carry out is a risk-based one using a multitude of input data types (see Section 5.5 for various recommended input data types). It is noteworthy that using more input data types further enhances the failure preemption capability. This is because of the fact that using more input data types increases both the depth and accuracy of identifying locations or areas that will become leaks or failures in the foreseeable future.

Figure 3.2 listed various inspection requirements that influence the accuracy and efficiency of any inspection activity:

- Inspection technique (e.g., UT wall thickness, radiography, and visual)
- Inspection type (e.g., time-based, consequence-based, or risk-based)
- Inspection frequency
- Inspection extent or area
- Susceptible features
- Data processing
- Data reliability
- Competency

The MCMM clearly highlights such inspection requirements; therefore, proper implementation of this particular corrosion management model can produce more accurate RBI scopes with greater corrosion failure preemption capabilities; hence lesser number of leaks per annum.

7.4.3 Corrosion Monitoring and Fluid Sampling Requirements

Similar to inspection review, a thorough and comprehensive review of the corrosion rate monitoring and fluid sampling systems can ensure that both accurate and reliable data are produced. This is achieved through carrying out an integrity review process on the following corrosion rate monitoring requirements (Figure 3.2):

- Corrosion rate monitoring method (i.e., probes or coupons)
- Location
- Orientation
- Service requirements
- Data processing and reliability
- Competency

A similar integrity review process has to be carried out on the following fluid sampling requirements (Figure 3.2):

- Method (i.e., online, manual, or both)
- Location
- Frequency
- Procedure
- Data processing and reliability
- Competency

The ultimate output of such integrity reviews and devising optimum corrosion rate monitoring and fluid sampling systems is the regular production of accurate and reliable corrosion rate and fluid data, leading to the creation of an improved asset integrity picture. Consequently, the incumbent corrosion and integrity issues are better identified, analyzed, quantified, and appreciated. This in turn facilitates improved integrity management decision-making and enhanced corrosion and integrity control, leading eventually to optimized AIM costs.

7.4.4 Management Requirements

The 12 Management Requirements (Chapter 4) play a significant role in the successful and efficient functioning of an AIMS. The ease of creating, updating, revising, and enhancing each of them, along with the fact that in many cases they need no or few resources (to create, update, and revise) renders them even more useful, practical, and desirable.

Often, small improvements associated with one or a few of them at no extra or negligible costs can make profound AIM improvements. Thus, carrying out a thorough integrity review process on them to make sure that they are present, up-to-date, useful, relevant, and effective is absolutely crucial.

The following 12 sections are dedicated to discussing the potential benefits of Management Requirements' applications.

7.4.4.1 Registers

Up-to-date and complete registers can significantly facilitate conducting an activity on one group, class, or type of equipment or parts thereof. Following are examples of such activities, where the success and efficiency of these activities depends on the thoroughness of the pertinent registers:

- Chemical treatment of the deadlegs (requiring a complete list or register of deadlegs)
- Failure risk assessment of all valves (requiring a complete list or register of valves)
- Determining the next year's piping inspections scope using an RBI software (requiring a complete piping list)
- Determining which insulated equipment will be inspected (requiring a complete list of insulated equipment)

Therefore, creating the relevant registers and maintaining them ensures proper and effective treatment, assessment, or selection of all or certain items within that register.

7.4.4.2 Strategies

A clear, practical, and well thought out strategy document can significantly improve the control and mitigation of certain deterioration mechanisms such as deadleg corrosion, corrosion under insulation (CUI) and microbially induced corrosion (MIC).

The efficient and successful mitigation of such deterioration mechanisms often requires a multistaged process involving various activities such as inspections, marking (certain locations on various) documents, chemical treatments, data analysis, and fluid sampling in a particular order. Therefore, the right strategy document can have a profound positive effect on the overall AIMS; thus, further optimizing the associated corrosion and integrity costs.

7.4.4.3 Procedures

Similar to strategies, procedures that are clear, relevant, and well thought out can have a significant positive effect on both the overall AIMS and effective corrosion mitigation across an asset. This is achieved by ensuring correct, timely, and effective carrying out of certain activities using an identical methodology.

One of the best procedural examples is the fluid sampling procedure. It can play a very critical role by ensuring that fluid samples are taken at the right time, from the designated locations, using the appropriate sampling techniques, etc. The right fluid sampling procedure, when fully implemented and adhered to, can markedly improve data accuracy and reliability associated with fluids' composition, characteristics, and corrosivity. Such information can in turn be used by the asset corrosion engineer to make better, more effective, and more relevant decisions; leading to improved corrosion mitigation and optimized corrosion and integrity costs.

7.4.4.4 Data Types and Processing (Data Management)

Effective and successful asset integrity management, to a great extent, depends on identifying the incumbent deterioration mechanisms followed by devising and implementing the appropriate integrity management measures. Both activities heavily rely on having access to various input data types and their analysis, trending, and processing.

Using a higher number of input data types and their timely and proper analysis greatly facilitates creating a clear and comprehensive integrity picture of the asset. This enables the asset corrosion engineer to better and more accurately identify the incumbent deterioration mechanisms; hence more appropriate and more effective integrity management measures are planned and implemented. The outcome is a more efficient AIMS associated with optimized corrosion and integrity costs.

7.4.4.5 Databases

Storing various input data in a neat, organized, transparent, and processable format improves their analysis, trending, conversion, and filtration. Any such improvements associated with data storage systems or databases would further enhance the asset corrosion engineer's knowledge and understanding about the asset's integrity condition, issues, and requirements. Hence, better and more informed integrity-related decisions can be made leading to more efficient AIMSs and optimized corrosion and integrity costs.

7.4.4.6 Documentation

Proper, thorough, and up-to-date documentation plays a crucial role in any AIMS. Document types such as process flow diagrams (PFDs), minutes of meetings (e.g., monthly corrosion meetings), annual asset integrity reports, regular lab reports, ad hoc technical notes, and relevant national and international standards all contribute to an overall asset integrity management system by creating an integrity picture of the asset concerned.

The better an asset's documents are, the clearer and more transparent the created integrity picture is. That means integrity issues are identified faster, and are better quantified and appreciated. Hence, better and more efficient rectifying or mitigation activities can be devised and implemented. The outcome is a more efficient AIMS and optimized integrity costs.

7.4.4.7 Asset Corrosion Management Strategy Document

Many of the encountered corrosion and integrity issues in the oil and gas industry can be traced back to the lack of or inadequate knowledge pertaining to an asset's basic integrity attributes:

- Existing materials and alloy systems
- Encountered and expected deterioration mechanisms
- Type and concentration of injected chemicals
- Type, location, and orientation of the existing corrosion rate monitoring devices
- Inspection techniques used
- High corrosion rate or risk areas

Simultaneously, adequate and thorough knowledge about the aforementioned asset integrity attributes further facilitates and improves AIM measures and activities implemented by the asset corrosion engineer and his colleagues. Thus, the creation and regular updating and maintenance of the asset corrosion management strategy document (also referred to as the asset corrosion control document) plays a crucial and indispensable role in enhancing the efficiency of an AIMS and optimizes the associated integrity and corrosion cots.

7.4.4.8 Corrosion Control Matrices

Corrosion control matrices (CCM) are one of the most crucial, useful, and indispensable corrosion management tools (as detailed in Section 4.9). Their timely creation and proper implementation and maintenance can have a profound positive influence on any existing AIMS. Ideally, CCMs should be determined at the asset's design stage and then immediately implemented after commissioning.

Full compliance with CCMs can reduce the repair and replacement costs. This is achieved through optimized chemical treatment and process parameters; thus, minimizing the incumbent deterioration rates.

7.4.4.9 Corrosion Key Performance Indicators (KPIs)

Corrosion KPIs (similar to CCMs) are considered as one of the most crucial and indispensable components of any corrosion management system. They are simultaneously a very useful tool for regularly monitoring the latest performances, and for highlighting any possible shortcomings and deficiencies associated with the incumbent corrosion management systems. Their inclusion in the monthly corrosion report significantly improves knowledge regarding the most important chemical and process parameters whose effective control within a designated range further minimizes the deterioration rates, while optimizing the associated corrosion and integrity costs.

7.4.4.10 Team Structure: Roles and Responsibilities

A complete team structure that is accompanied by detailed and clear roles and responsibilities (or job descriptions) imparts further clarity and transparency into any AIMS. Thus, there is lesser likelihood that a particular integrity-related task or an integrity management measure is neglected, forgotten, or abandoned.

One of the most important components of this is the *absolute requirement* for the existence of the corrosion engineering position within the team structure; this is especially critical for those oil and gas assets where one of the main integrity threats is corrosion. The presence of a competent asset corrosion engineer can significantly improve the efficiency of an AIMS through the implementation of more effective corrosion engineering and corrosion management principles.

7.4.4.11 Competency

Competency and communication are considered as two of the most important and influential non-CE–based integrity management measures that have the greatest effect on the performance and efficiency of an AIMS. Many of the worst corrosion integrity issues encountered by the author in the oil and gas industry around the world were due to the total lack of or inadequate, competency levels associated with the pertinent disciplines within the incumbent AIMSs.

On the other hand, some of the best, i.e., most successful and most efficient, AIMSs ever encountered internationally by the author, owed their success and effectiveness to highly competent individuals responsible for devising and implementing such integrity management systems. Adequate competency levels can further enhance an organization's failure preemption capability significantly (see Section 8.2.5) and markedly optimize the pertaining corrosion and integrity costs.

7.4.4.12 Communication

As mentioned previously, competency and communication are considered as two of the most crucial and influential non-CE–based integrity management measures. Adequate and timely communication among relevant disciplines within the AIM team and beyond (where and when necessary) has a tremendous effect on any incumbent asset integrity management system. Improved communication can enhance general knowledge about different critical aspects of an asset (such as inspection findings, corrosion rate monitoring data, anomaly and leak data, and so on) leading to better and more informed integrity decisions being made. Consequently, the incumbent deterioration rates are mitigated more efficiently and the associated corrosion and integrity costs are optimized gradually and over time as the level, content, timing, and quality of communication improves.

Last, but not the least, communication shortcomings and deficiencies can be improved–in most cases at no or very little costs–leading to potential and profound enhancements in the efficiency of the incumbent AIMS.

7.4.5 Failure Risk Assessment (FRA) Requirements

A failure risk assessment approach ensures that all inspections carried out are risk-based. Implementing an inspection scope that has been created through using the criterion of risk further improves the failure preemption capability of an organization. Thus, corrosion failures become less frequent, and accordingly postfailure corrosion costs decrease (see Section 7.2.6).

A risk-based approach toward inspections can also optimize repair and maintenance activities. By providing more detailed information about high failure risk areas that might develop into failures in the foreseeable future, better and more effective decision-making on such matters is facilitated.

7.5 Conclusions

Proper and timely implementation of the MCMM can bring about many potential benefits. An important component of this model is the integrity review process through which various integrity management measures can be thoroughly reviewed. Thereafter, if needed, such integrity management measures can be revised, updated, and improved leading to a more efficient AIMS.

Below are some of the most critical benefits gained through corrosion management process implementation:

1. Enhanced corrosion failure preemption capability
2. Optimized corrosion and integrity costs
3. Improved identification and quantification of the incumbent deterioration mechanisms
4. Enhanced competency and communication
5. Increased asset uptime
6. Improved personnel safety and environmental protection
7. Risk-based inspections

8 Observed Shortcomings Associated with the Corrosion Management Implementation Process

8.1 Background

As was mentioned in Section 3.13, the corrosion management implementation process is composed of the following three phases:

1. Implementation Phase I: Pre-Implementation
2. Implementation Phase II: Implementation
3. Implementation Phase III: Post-Implementation

Accordingly, any possible issue or shortcoming pertaining to the corrosion management implementation process can be associated with any of these phases. Therefore, implementation issues have similarly been categorized into three separate groups as explained in the following sections.

The foremost implementation issue is still, however, not implementing or carrying out the corrosion management process at all. That is, unfortunately, a common and widespread practice in numerous industries (where corrosion is a major integrity threat), including oil and gas.

8.2 Shortcomings Associated with the Pre-Implementation Phase

The following five sections focus on implementation issues associated with the Pre-Implementation Phase.

8.2.1 Issues Associated with the Scope of Work

Any shortcomings associated with the scope, such as uncertainties, ambiguities, or erroneous instructions, specifications, requirements, and prerequisites would produce a defective, inefficient, incomplete, or erroneous corrosion management implementation process and outcome. Typical shortcomings or issues associated with the scope of work can include, but are not limited to the following issues:

- Unclear or vague physical boundaries for the work or project
- Unclear or vague statements or specifications regarding systems' or equipment's inclusion or exclusion
- Poor and/or ambiguous description of the work to be carried out
- Not basing the work to be carried out on any known documents, standards, or methodology

Any other similar poor and inadequate information, instructions, clarifications, or drawings can also further render the scope of work *less* transparent, *less* understandable, and *less* practical. Thus, it would adversely affect the entire corrosion management implementation process and its outcome.

8.2.2 Lack of, or Inadequate, Management Support

One of the crucial prerequisites for any successful corrosion management implementation process is the allocation of the appropriate and required resources by the management. Any shortcomings in the quantity or quality of these required resources can lead to an incomplete and ineffective implementation process. Therefore, it is crucial that adequate corrosion management awareness training is provided for the relevant senior management, so they would realize and fully appreciate the significance of the corrosion management implementation process and its potential benefits. Otherwise, there is no point in trying to proceed with the implementation process without prior provision of corrosion management awareness training for the management.

Several field observations have demonstrated the utter failure of the intended implementation process when the management was not engaged from the beginning.

8.2.3 Lack of, or Inadequate, Site Personnel Competency

The implementation process is always carried out with the help, support, and cooperation of the pertinent site personnel. Therefore, prior to any implementation, adequate corrosion management training should be provided to the pertinent site personnel, so they would know what the corrosion management concept is, how it is implemented, what its components are, etc. Any possible shortcomings associated with site personnel corrosion management training can simply doom the entire implementation process.

Past field experience has shown that when prior training is not provided for the relevant personnel, the implementation process would not go far and would achieve very little, if at all.

8.2.4 Lack of, or Unreliable, Input Data

One of the major issues associated with some implementation processes is the lack of the required input data or providing unreliable data and information. Lack of certain crucial data can almost totally stall the implementation process, while the provision of unreliable data can lead into the generation of erroneous outputs (e.g., corrosion rates and risk levels).

Therefore, prior to the beginning of any implementation process, one has to verify various data availability and reliability to ensure that there exists adequate reliable input data to enable the implementation phase to proceed.

8.2.5 Inadequate Team Structure and Job Descriptions

The corrosion management implementation process is a team exercise–every participating individual should be fully aware of their relevant responsibility within the team. Any possible lack of, or inadequate, team structure and job descriptions would often lead to task or activity overlap, or partial or total oversight of certain responsibilities, leading to poor or incomplete implementation processes.

8.3 Shortcomings Associated with the Implementation Phase

The following four sections focus on implementation issues associated with the Implementation Phase.

8.3.1 Not Implementing the Corrosion Management Process

The most prevalent issue or shortcoming observed across the oil and gas industry has been not implementing the corrosion management process itself; in particular, throughout an asset's post-commissioning or operations phase. The main culprit behind such shortcoming is that there are some individuals (within the industry) who still regard the corrosion management concept as a mere synonym for the corrosion engineering discipline. Furthermore, by assuming that the crucial corrosion engineering requirements have already been fulfilled during the asset's design and construction phases, they mainly focus on meeting and maintaining target production levels during the asset's operations phase; at almost any costs, or more precisely at the cost of an asset's deteriorating integrity condition without doing or devising much preemptive or effective mitigation measures.

Such mentality and approach toward asset integrity management leads to increased corrosion and integrity leaks and failures with increasing numbers and severity as the asset ages.

8.3.2 Ignoring Non-CE-Based Integrity Management Measures

Reviews of various corrosion management implementation projects did reveal that only CE-based integrity management measures were covered during the conducted integrity reviews and the non-CE-based ones were totally ignored. Thus, throughout such implementation projects, the implementation team failed to highlight the faulty, out-of-date, ineffective, or even obsolete non-CE-based integrity management measures that were the main root causes of the incumbent integrity and corrosion issues.

Unfortunately, this is still a common practice across various assets, where only CE-based integrity management measures are taken into account. Such implementation projects fail, in most cases, to identify all the improvement opportunities because of their ignoring the non-CE-based integrity management measures.

8.3.3 Ignoring Management Requirements

In spite of the fact that the management requirements are considered the most crucial of all the non-CE-based integrity management measures, nevertheless, very rarely they are included in the associated integrity review and implementation processes. This (mostly) inadvertent exclusion of the management requirements can be regarded as the main culprit behind failing, unsuccessful, and ineffective corrosion management implementation processes observed or studied worldwide.

Within the management requirements, communication and competency items are simultaneously regarded as the most important of the management requirements. Any implementation process should make every effort to include them as well. However, similarly these two crucial items are very seldom covered in the observed or studied implementation processes.

8.3.4 Focusing Solely on the Creation of Corrosion Loops

The author's field visits and corrosion management audits have demonstrated that in many corrosion management implementation cases, the implementation process was limited to the creation of corrosion loops and the carrying out of the failure risk assessment (FRA) process (or RBI) for such loops. What still exacerbated the situation was that those who had created the corrosion loops had used erroneous criteria and definitions to create the loops. Such discrepancies and mistakes had, in later years, culminated in increasing integrity and corrosion failures because in some instances, very high corrosion rate areas were placed in the same loop with low corrosion rate areas and the associated inspection interval was adjusted with regards to the latter.

This rather widespread idea that the corrosion management implementation process is merely synonymous with the creation of corrosion loops should be scrupulously avoided.

8.3.5 Focusing Solely on the Wall Thickness Inspection Data

To carry out a comprehensive and reliable failure risk assessment (FRA) process one can use various data types:

- Design
- Operation
- Process
- Chemical treatment
- Corrosion rate monitoring and sampling
- Inspection
- Cathodic protection
- Coating
- Repairs and replacements
- Anomalies
- Leaks

However, the author has observed on a very regular basis that in many assets—mainly to save time—the available wall thickness data was referred to and used as the *sole input* for the FRA process. This erroneous approach often leads to unrealistic risk levels and wrong inspection intervals. Unfortunately, the ultimate outcome is the gradual increase of the integrity and corrosion issues, along with growing integrity management costs, due to the above erroneous (FRA) practice.

8.4 Shortcomings Associated with the Post-Implementation Phase

The following four sections focus on implementation issues associated with the Post-Implementation Phase.

8.4.1 Not Implementing Recommendations

One of the main outcomes of the integrity review process in particular, and the corrosion management implementation process in general, is the creation of recommendations that pertain to both CE-based and non-CE-based integrity management measures. If such measures do not exist already, then they need to be created; and if they already exist, they might require revising, updating, or enhancing. Therefore, implementing such recommendations is expected to improve both the effectiveness and the efficiency of the incumbent asset integrity and corrosion management systems.

Studying various site reports associated with the post-implementation phase, however, revealed that the majority of the recommendations were not implemented–that is, the existing corrosion management system was not improved in any way. Consequently, integrity and corrosion issues had continued to grow both in number and size afterwards.

8.4.2 Cessation of Performance Monitoring

A misunderstanding exists that the process of performance monitoring is mainly limited to the time when the integrity review process is ongoing. Based on this erroneous approach, once the integrity review process is completed, then the performance monitoring process is stopped until the next round of integrity reviews in a few years' time. With this incorrect approach, vital data–such as corrosion rate monitoring and fluid sampling data throughout the asset operations phase–is not available.

8.4.3 Cessation of Effectiveness Assessment

Similar to the shortcoming (for the performance monitoring process) described in Section 8.4.2, in many assets, the process of assessing the effectiveness of various integrity management measures is limited to the time when the integrity review process is ongoing and the performance monitoring is being carried out. Once the integrity review process is over, performance monitoring and the ensuing effectiveness assessment processes are ceased until the next round of integrity reviews.

This practice of ceasing the performance monitoring and effectiveness assessment processes virtually brings the whole corrosion management implementation process to a standstill and should be absolutely avoided.

8.4.4 Cessation of Communication Between Two Parties

Another erroneous practice throughout the post-implementation process is the total cessation of communication between the individual or the team (i.e., the consultant engineers or the consulting company) who carried out or facilitated the corrosion management implementation process and the site personnel. The author's experience has shown time and time again that because the two parties are not communicating with each other, then the consultant engineer or consulting company cannot have access to any data (e.g., corrosion rate and corrosion KPI data), which can tell them how the site personnel are performing. Furthermore, the consulting engineers cannot provide any assistance, consultancy, and support when the site personnel face implementation issues or when they require assistance in implementing the recommendations. This lack of communication is one of the most prevalent issues during the post-implementation phase with often detrimental (integrity) consequences for the asset.

Corrosion Management Implementation Options

9.1 Background

As previously discussed in Section 3.13, any corrosion management implementation process is composed of these three phases:

- The Pre-Implementation Phase
- The Implementation Phase
- The Post-Implementation Phase

This chapter exclusively discusses the Implementation Phase and the recommended options for carrying out this particular phase. There are three recommended implementation options, one of which is further divided into two subcategories:

1. Complete Implementation Option
2. Smart Implementation Option
   a. Focuses on the 12 management requirements
   b. Focuses on six particular integrity management measures
3. Targeted or Tactical Implementation

In general, it is either the available resources, or the immediate integrity requirements of the asset, or a combination of both that often determine which of the above options will be taken. It is highly recommended to involve the corrosion management consultant (who is supposed to provide the pertinent advice and consultancy throughout the implementation process) in the decision-making process, so the most appropriate, relevant, and efficient option is selected.

9.2 Complete Implementation

In this option, the existing CE-based and non-CE-based integrity management measures (Figure 3.2) are taken into account and the corrosion management process is carried out. *This option is the most comprehensive and thorough implementation option requiring the greatest resources and input data among all implementation options.* Accordingly, it can take significantly longer to complete, however the resultant integrity picture that it creates (regarding the incumbent integrity condition of the asset) is the most comprehensive one. Therefore, more relevant, up-to-date, and effective integrity solutions can be devised and then implemented.

In cases where the required resources (i.e., various disciplines, budget, and time) are not available to fully implement this option quantitatively, this option can be carried out in a more qualitative-based manner. In this way, creating the full integrity picture of the asset, albeit on a more qualitative base rather than quantitative, is still possible.

9.3 Smart Implementation

This option focuses on the implementation of all, or some, of the most important management requirements (Chapter 4). This implementation option is further divided into two subcategories:

1. Subcategory 1: Focusing on the 12 management requirements (see Section 9.3.1)
2. Subcategory 2: Focusing on six basic integrity management measures (see Section 9.3.2)

Both subcategories can be implemented very quickly; and they require the least amount of resources compared to the other two implementation options (Section 9.1). That is why this option is called the "smart" option.

9.3.1 Focusing on the 12 Management Requirements

A large number of corrosion issues encountered in the oil and gas industry are deeply routed in one or a combination of the 12 management requirements (Chapter 4). Therefore, by creating, updating, revising, correcting, completing, or enhancing the relevant management requirements an asset corrosion engineer can make profound asset integrity improvements such as the following, often at no or very little costs:

- **Procedures:** Creating a clear, precise, and transparent fluid sampling procedure (in a matter of an hour or less for asset corrosion engineers who are adequately familiar with their asset and its integrity condition) would significantly improve the consistency and accuracy of the fluid samples taken. The lab reports (which are produced after analyzing such fluid samples) would be more relevant and accurate thereby helping to produce a clearer, more consistent, and more accurate integrity picture of the asset or parts. Thus, more efficient and relevant integrity management decisions can be made, resulting in an enhanced AIMS.

- **Communication:** Establishing or initiating regular (e.g., monthly) corrosion meetings among the pertinent disciplines and teams (at virtually no extra costs) would further facilitate data and information dissemination. Such improved data and information sharing and exchange enhances general understanding about the incumbent integrity condition of the asset, which leads to better and more effective (integrity) decision-making, rectifying activities, and more realistic deadlines.
- **Registers or Databases:** Creating and maintaining a leak register would significantly facilitate data gathering, analysis, trending, and filtration associated with identified corrosion leaks. Such leak data processing and analysis can markedly facilitate the ongoing failure investigation projects or tasks by making it easier to determine the pertinent (leak) root causes more efficiently. Consequently, more relevant, effective, and bespoke rectifying actions can be devised and implemented, leading to an enhanced and more efficient AIMS.

9.3.2 Focusing on Six Basic Integrity Management Measures

Various corrosion management implementation and audit assignments associated with both the hydrocarbon and the non-hydrocarbon industries have highlighted the fact that there are six non-CE–based integrity management measures that play a more profound role throughout an asset lifecycle (from design to decommissioning) in relation to the other non-CE–based integrity management measures.

That is, creating, improving, updating, revising, and completing so few non-CE–based integrity management measures can bring about various and significant asset integrity improvements in very little time and at (often) very low costs. Simultaneously, based on extensive feedback the author received from hundreds of delegates attending corrosion management training courses in the past few years, the same six non-CE–based integrity management measures were underlined as the most indispensable and influential ones that–at a minimum–should be present throughout any corrosion management implementation process.

The six non-CE–based integrity management measures listed below were covered in detail in Chapters 4 and 6:

1. Communication
2. Competency
3. The Presence of an Asset Corrosion Engineer Position
4. Risk-based Inspections
5. Corrosion Control Matrices
6. Corrosion Key Performance Indicators

9.4 Targeted or Tactical Implementation

Sometimes the asset in question is suffering from a particular deterioration mechanism (e.g., severe and widespread bacterial contamination) or acute and severe corrosion rates associated with certain systems, areas, or equipment types that need the most urgent attention, resource allocation, and rectification activities. The asset corrosion engineer begins implementing the corrosion management process by first targeting the high priority integrity issues, systems, or areas. These high priorities have to be rectified before moving to other areas with lower integrity priorities. Alternatively, once the targeted area or system has been dealt with and the integrity issue rectified or mitigated, then one can resort to other available options to continue the implementation process.

Through this option, the most severe or prevalent deterioration mechanisms can be targeted and mitigated in a shorter time with simultaneously greater dedication of available resources; resulting in a very efficient corrosion management implementation process.

9.5 Conclusions

There are different corrosion management implementation options that can be chosen based on an asset's age, integrity condition and requirements, the available resources, and lastly the operator's preferences. Each individual implementation option is associated with particular attributes:

- Thoroughness
- Speed
- Low costs
- Efficiency
- Ease of application

Here are the main advantages associated with each implementation option.

- Option 1 (Complete Implementation): the most thorough option; requires more input data; requires longer time to complete; the most expensive
- Option 2 (Smart Implementation Option)
  - 2a (Focus on the 12 management requirements): one of the most effective options; fast; one of the least expensive
  - 2b (Focus on six particular integrity management measures): has minimum implementation requirements; fast; the least expensive
- Option 3 (Targeted or Tactical Implementation): one of the most efficient options; better use of available resources

References

1. Koch, G.H., M.P.H. Brongers, N.G. Thompson, Y.P. Virmani, and J.H. Payer, "Corrosion Costs and Preventive Strategies in the United States," Federal Highway Administration, publication no. FHWA-RD-01-156, 2002 (Houston, TX: NACE International).
2. Capcis Limited, "Review of Corrosion Management for Offshore Oil and Gas Processing" (Suffolk, United Kingdom: Health and Safety Executive, 2001).
3. HSE, Asset Integrity Programme, Key Programme Three (KP3), "Report by the Offshore Division of HSE Hazardous Installations Directorate" (Suffolk, United Kingdom: Health and Safety Executive, 2007).
4. Morshed, A., *An Introduction to Asset Corrosion Management in the Oil and Gas Industry* (Houston, TX: NACE International, 2012).
5. Energy Institute, *Guidance for Corrosion Management in Oil and Gas Production and Processing* (London, United Kingdom: Energy Institute, 2008).
6. Energy Institute, *Guidance for Corrosion Management in Oil and Gas Production and Processing*, 2nd ed. (London, United Kingdom: Energy Institute, 2019).
7. API RP 580, "Risk-Based Inspection," 3rd ed. (Washington, DC: American Petroleum Institute, 2016).
8. API RP 570, "Piping Inspection Code: In-Service Inspection, Rating, Repair and Alteration of Piping Systems," 4th ed. (Washington, DC: American Petroleum Institute, 2016).
9. API RP 510, "Pressure Vessel Inspection Code: In-Service Inspection, Rating, Repair and Alteration," 3rd ed. (Washington, DC: American Petroleum Institute, 2014).
10. API RP 653, "Tank Inspection, Repair, Alteration and Reconstruction," 5th ed. (Washington, DC: American Petroleum Institute, 2014).

Corrosion Management Case Studies

Introduction

This appendix contains case studies that further illustrate how the corrosion management process and applications can be used to

- Better identify and determine (both the CE–based and the non-CE–based) shortcomings that often cause the pertinent corrosion and integrity failures and leaks
- Illustrate how some of the existing, acute, and costly corrosion issues have their roots in sometimes very trivial and easy-to-rectify misses, oversights, shortcomings and wrongdoings
- Demonstrate how easily some rectifying actions can be devised (at no or very little cost in many cases) to remove or improve the situation once the root cause(s) has been identified
- Optimize various pertinent integrity costs over time

The case studies pertain to actual corrosion management implementation projects, audits, or asset integrity reviews carried out by the author. The assets covered included offshore production platforms, subsea and overland pipelines, onshore terminals, and downstream processing facilities.

The case studies have the same structure with the following components.

Overview: A brief summary of the asset and the incumbent corrosion situation.

Observation: The main finding(s) of the Corrosion Management Team during their visit/integrity review.

Culprit: The main corrosion management shortcoming(s) and/or uncompliance(s) identified.

Outcome: The most significant repercussion(s) of the identified corrosion management shortcoming(s).

Corrosion Management Recommendations: The main rectifying action(s) suggested by the Corrosion Management Team (whether it was implemented by the asset owner or not).

Note: To protect the location and the true identity of the asset owners/operators and the contractors, the physical description (i.e., the location, number of trains, age, or any other physical characteristics) of the assets covered in the following case studies have been slightly modified.

Case Study 1: An Offshore Seawater Injection System

Overview: An offshore production platform also has a seawater injection system. This system suffers from frequent leaks, and therefore requires regular repair and maintenance shutdowns. Because of escalating integrity costs, the operator requests a corrosion management audit to be carried out.

Observation: During the integrity review it became evident that routine integrity management activities (for a typical seawater injection system) were not carried out. Such activities should have included

- Continuous oxygen scavenging
- Frequent biociding
- Monitoring the residual concentration of chlorine upstream of the deaeration system
- Monitoring the residual concentration of oxygen downstream of the deaeration system
- Bacterial enumeration

Furthermore, the pertinent personnel were unaware of the

- Correct injection rates for the oxygen scavenger and the biocide chemicals
- Residual concentration limits of chlorine (upstream of deaeration) and dissolved oxygen (downstream of the deaeration)
- Acceptable bacterial levels at different sections of the seawater injection system

Culprit: The CCM Document was not updated over the years and was completely obsolete. Therefore, the personnel did not trust the numerical values listed there for the above activities; consequently, they stopped carrying them out gradually, leading to the current situation (i.e., lack of proper up-to-date documentation and competency issues).

Outcome: Frequent leaks and operations shutdowns due to corrosion, leading to a costly repair and maintenance program.

Corrosion Management Recommendations: Updating the out-of-date corrosion control matrices (CCM) document, along with providing the necessary training on the integrity management of seawater injection systems for the relevant personnel.

Case Study 2: An Offshore Production Platform

Overview: A mature offshore production platform, whose topside production process piping and equipment are mostly made from carbon steel, has been in operation for more than 20 years. A significant number of deadlegs exist onboard the platform; therefore, the operator requests an integrity review to be carried out.

Observation: The integrity review revealed that many operational and physical deadlegs existed, which, because of the increased water cut and age of the asset, posed a high integrity threat. The operator was apparently fully aware of the situation, but continued to ignore the worsening integrity situation with complacency. The operator refused to create a "Deadleg Strategy" to deal with the situation, quoting the costs associated with compiling and implementing such strategy as the main hindrance.

Culprit: The asset operator's complacency that everything was under control, plus the operator's lack of appreciating the true nature, extent, and severity of the deadleg situation onboard their asset–in spite of the warnings issued by the contractor's Corrosion Management (CM) Team (i.e., lack of proper documentation and competency issues)–were at fault.

Outcome: Weeks after the warning issued by the CM Team, the process piping had a hydrocarbon leak in a deadleg section. Consequently, the whole asset was shut down and a costly deadleg maintenance program was urgently initiated. The incident also involved Health and Safety Executive (HSE) and was a huge reputation fiasco for the asset owner, asset operator, and pertinent management.

Corrosion Management Recommendations: The CM Team recommendation was to identify all the deadlegs and carry out a risk assessment to identify the ones with the highest failure risk and lowest remaining lives. Thereafter, a strategy was to be created to deal with the deadlegs in a way (e.g., physical elimination, chemical treatment, and/or more frequent inspections) to continuously reduce the failure risk posed by them.

Case Study 3: An Onshore Seawater Cooling System

Overview: A seawater injection system has been constructed to use Persian Gulf water as a cooling medium for some industries farther downstream. However, somehow no deaeration or oxygen scavenging has been associated with the asset's design, whose main material is carbon steel. A short time after commissioning, the asset starts suffering from multiple pinhole leaks. After a while and because of the worsening situation, an oxygen scavenging system is retrofitted, however, the leak problem continues. Therefore, the operator requests an integrity review to be carried out.

Observation: The integrity review was carried out and it identified that no monitoring system existed to monitor the concentration of the residual oxygen. In other words, the performance of the retrofitted oxygen scavenging system was not monitored in any way to determine how successful or effective it had been.

Culprit: Poor design considerations followed by poor operation and maintenance, and failing to identify the integrity threats at both stages (i.e., lack of competency at design and operation levels).

Outcome: Frequent operation shutdowns due to multiple leaks not long after commissioning, followed by retrofitting a deaeration system that lacked a performance monitoring system.

Corrosion Management Recommendations: The main recommendation was to create a performance monitoring system to measure the performance of the deaeration system to determine if the dissolved oxygen concentration downstream of the deaeration system was in compliance with the required treated seawater specification. The recommended performance monitoring system should be composed of

- An online oxygen monitoring system
- A manual oxygen monitoring system
- A corrosion rate monitoring system (e.g., coupons and probes)

Case Study 4: An Offshore Production Platform

Overview: An offshore production platform produces sweet wet gas through two trains and the platform process piping is made of carbon steel. Regular wall thickness inspections are carried out and sent to the office for both data trending and analysis. At one point, the office orders an operation shutdown in one of the production trains to replace a few carbon steel pipes because of corrosion at a significant cost.

Observation: An integrity review revealed that the wall thickness data sent to the office was not reviewed, analyzed, or trended at all. No corrosion rates or remaining life values were calculated either. Because nobody in the office had any idea about the prevailing corrosion rates at different parts of the platform, the replacement was just requested without any real reason to do so. To be on the safe side, however, they decided to act conservatively and change out any pieces that they *assumed* were suffering from higher corrosion rates.

Culprit: Total lack of wall thickness data review, trending, and analysis, which resulted in the office's responsible corrosion engineer not knowing what was happening in terms of corrosion in the above offshore platform.

Outcome: The above conservative approach to change out a few pipe pieces was very costly because it caused a 50% drop in production rates. Other costs suffered were the associated labor cost and the cost of the new piping spools. Later inspection of the replaced spools revealed that their actual remaining thickness was still near the nominal thickness levels!

Corrosion Management Recommendations: The CM Team requested that all wall thickness inspection data be fully processed and any further equipment replacements are supported by pertinent integrity assessments, estimated corrosion rates, and remaining life values. In addition, because the actual remaining thickness of the removed piping spools were still near the nominal thickness levels (something that had totally been missed by the inspection team) it was decided that the inspection team had to undergo a refresher inspection training course to improve their technical inspection skills and conduct.

Case Study 5: An Onshore Gas Terminal

Overview: An onshore gas terminal receives gas from offshore and treats it across its multiple processing trains. The bulk of the processing piping and equipment is made of carbon steel that is inspected based on an RBI program. Several field inspectors carry out the required inspections, and the wall thickness data are passed on to the asset's corrosion engineer for processing.

Observation: An integrity review was carried out that also included reviewing the available wall thickness inspection data for the past few years. Such a data review showed that significant portions of the provided wall thickness data varied in the same fashion over time. That is, one year they reduced and the next year the wall thicknesses increased. This pattern was repeated across the whole database. This led the CM Team to review the qualifications of the incumbent inspection team and the finding was that they were not qualified at all to carry out ultrasonic testing (UT) of wall thickness inspections.

Culprit: Lack of competency screening and control across various disciplines working for the asset; in particular, total lack of competency (due to poor initial training) associated with the UT wall thickness inspection team.

Outcome: The bulk of the wall thickness data produced over the years was not reliable and could not be used to determine or estimate the pertinent corrosion rates and remaining life values. In other words, no integrity assessments could have been carried out using such data.

Corrosion Management Recommendations: The CM Team's main recommendation was to either provide a comprehensive training course for the incumbent UT wall thickness inspection team or replace them with a fully qualified and better equipped replacement.

Case Study 6: Five Offshore Production Platforms

Overview: The integrity of five offshore platforms is managed by an onshore integrity management team including corrosion engineers, inspection engineers, chemical treatment personnel, and personnel who oversee the corrosion-monitoring systems. All five platforms suffer from various corrosion issues and leaks on a regular basis. The integrity management team holds monthly corrosion meetings to discuss and tackle the various corrosion and integrity issues pertaining to these five offshore production platforms.

Observation: During the integrity review it became clear that the various communications (both written and oral) taken place during the monthly corrosion meetings were neither effective, nor adequate. There was no single monthly report because each different group (e.g., the corrosion engineers, inspection engineers, and the chemical treatment responsible persons) brought their own report (or sometimes no reports) to the meetings. While various discussions took place, no rectifying actions or solutions were agreed upon and nobody was tasked with carrying out such activities.

Culprit: Extremely poor written and oral communication among various pertinent disciplines, which resulted in them not being able to determine the real cause(s) of corrosion across five offshore assets. This prevented them from planning and carrying out the required remedial actions to rectify or improve the worsening situation.

Outcome: Worsening integrity situation of all five platforms over time due to an almost complete breakdown in communication between the pertinent disciplines. Shutdowns were regular and the associated costs increased steadily.

Corrosion Management Recommendations: The main recommendation was to appoint one single person responsible for issuing the monthly corrosion report and distribute it a few days prior to each meeting to give adequate time for each participant to study the report and prepare. In addition, that person had to assign actions to the relevant individuals who would rectify or improve the existing corrosion issues. Finally, that person had to take the minutes of each meeting for future reference.

Case Study 7: A Subsea Gas Transport Pipeline

Overview: A subsea transport pipeline made of carbon steel is used to transfer sweet wet gas from an offshore platform to an onshore gas terminal. A corrosion inhibitor has to be injected into the pipeline inlet to protect it against corrosion. However, during an integrity review it became evident that no corrosion inhibitor had been injected for a while and the situation was going to continue for a few more months because of the lack of the chemical.

Observation: The integrity review revealed that there was not a well-defined job description to make somebody fully responsible for monitoring the stock levels of various chemicals–including the corrosion inhibitor. Therefore, it was only when the stock levels were almost fully depleted that somebody noticed they had to act immediately and order the chemical; however, the lead-time was rather long so the operator did not have any corrosion inhibitor to inject into the pipeline for several months.

Culprit: Roles and responsibilities were not clearly defined; therefore, nobody was clearly in charge of monitoring the chemical stock levels to order the required chemicals on time.

Outcome: The pipeline suffered significantly higher corrosion rates during the period when the corrosion inhibitor was not injected.

Corrosion Management Recommendations: The main CM Team recommendation was to introduce a new item in the Corrosion Control Matrices (CCM) Document to make the asset lab chemist responsible for regularly monitoring the chemicals' stock levels and order new chemicals when required. Furthermore, it was recommended to review all the existing job descriptions to update and clarify the relevant roles and responsibilities.

Case Study 8: An Offshore Production Platform

Overview: A mature offshore platform has been in operation for more than 25 years and is composed of oil, gas, and water injection systems. The process piping and equipment have been regularly inspected by an integrity management contractor who uses an inspection software program. The software allegedly is an RBI software and the inspection scope it produces for every system on board the offshore platform is fully risk-based. However, a few leaks due to corrosion, along with some near misses are identified; which put the performance of the software into question.

Observation: A very detailed integrity review was conducted to find out why the RBI software had missed various points with low remaining life values in past inspection campaigns. After a conscientious review of the scope of past inspections and the software program itself, it became evident that the software had various bugs and faults that prevented it from creating a fully risk-based inspection. It was also determined that by revising the scope of the annual inspection–which was fully risk-based–there were 50% less features to inspect. This was a significant reduction in the extent of annual inspections associated with similar cost reductions.

Culprit: Faulty inspection software, along with complacent software operators who had apparently not realized that the software that had been used for several years and was allegedly an RBI software, could not produce a real or fully risk-based inspection scope (i.e., lack of competency among the software developers and operators).

Outcome: Huge annual overspend on inspection activities because the produced annual inspection scopes were around 50% bigger than the real risk-based one.

Corrosion Management Recommendations: The main CM Team recommendation was to either fix the existing software bugs to enable it to produce real RBI scopes or to replace it with another software with risk-based capabilities. Furthermore, it was recommended to produce a new document dedicated to describing the RBI software and its components to facilitate the RBI methodology review in the future.

Case Study 9: Three Subsea Pipelines

Overview: There are three subsea pipelines, each connected to a subsea wellhead and they carry a sweet wet gas. There are injection points downstream of the choke valve at the inlet of each pipeline, and the corrosion inhibitor is injected continuously into all three pipelines, which are entirely made of carbon steel. Earlier wall thickness inspections revealed heavy metal losses associated with the first spool of one of the pipelines shortly after the spool's commissioning, which resulted in its replacement at a huge cost. Later inspections of the newly installed spool and the first spools of the other two pipelines also identify heavy localized corrosion. At this point the operator requests an integrity management contractor to carry out an integrity review on the pipelines to determine the likely cause of such internal corrosion.

Observation: The CM Team reviewed all the available pipeline data (i.e., design, process, chemical treatment, and past inspections) and noticed that such severe corrosion was only associated with the first pipeline spool immediately downstream of each choke valve and that the rest of the three pipelines were suffering from negligible corrosion. Thus, it was deduced that the most likely issue was that the tip of the three injection quills were damaged because of sand production; hence, they could not properly and completely mix or disperse the injected chemical with the produced wet gas within the length of the first spool.

Culprit: When this was presented to the pipeline owner, the owner said that no quills were installed from day one (i.e., lack of competency and ignoring own company standard to install quills at such locations).

Outcome: Huge replacement costs associated with replacing the first spool for one of the subsea pipelines, along with costly regular intelligent pigging to monitor the rate of corrosion for all three of the first spools and beyond.

Corrosion Management Recommendations: The pipeline owner mentioned that retrofitting injection quills was totally unacceptable because of the huge costs involved. Therefore, the CM Team recommendation was to carry out an integrity review for the other pipelines operated and/or owned by the same company in that region to identify whether there were similar cases or not.

Case Study 10: A Subsea Pipeline

Overview: Several subsea wells feed into a subsea manifold via their associated corrosion resistant alloy (CRA) flowlines. From the manifold, a single carbon steel pipeline carries the sweet wet gas to an offshore production platform for dehydration and further processing. Corrosion inhibitor is mixed with methanol, and then the mixture is continuously injected at the pipeline inlet at the manifold. Because the pipeline had been in operation for a very long time, the pipeline owner asked an integrity management contractor to carry out an integrity assessment and estimate the remaining life of the pipeline.

Observation: All the relevant pipeline data (e.g., design, process, chemical injection, corrosion monitoring, and past inspections) were acquired and an integrity review was carried out for the pipeline. The integrity review revealed that the incumbent corrosion inhibitor had undergone extensive lab testing prior to its selection. Based on the provided lab report the maximum required concentration of the corrosion inhibitor at the pipeline inlet was 20 ppm. The provided chemical injection data, however, indicated the current injection rate was actually 4,000 ppm. That is, the corrosion inhibitor was overdosed 200 times above its required limit!

Culprit: When the findings were presented to the pipeline owner, the explanation they provided was that over the years they had used some inexperienced individuals who were responsible for determining the required injection rate, using a spreadsheet program and that these young individuals had made simple calculation mistakes, gradually increasing the injection rate from 20 ppm to 4,000 ppm over a few years.

Outcome: Because of the huge overdose of the corrosion inhibitor over the years, the associated chemical costs were significantly higher (approximately 200 times over at the time of the review). This was the worst overdose case ever encountered by the author, caused by lack of competence among those responsible for managing the pipeline's chemical injection).

Corrosion Management Recommendations: The main recommendation was to reduce the injection rate to the required levels. It was also recommended to carry out some lab testing at 4,000 ppm to determine whether such high corrosion inhibitor concentration was detrimental to a carbon steel substrate or not. This was to identify whether or not such huge chemical overdoses had any adverse corrosion effect on the pipeline interior.

Case Study 11: An Onshore Gas Terminal

Overview: An onshore gas terminal receives sour wet gas from a number of gas wells. The terminal's main function is to dehydrate the gas and then send the dry gas for further processing to an onshore refinery. There have been many leaks associated with different equipment of the gas terminal, downstream of the dehydration system. A corrosion management team is dispatched to find out the root cause of such corrosion issues.

Observation: During the integrity review it was revealed that the dehydration system had been shut down for a number of years and the sour wet gas would flow through the downstream sections (of the dehydration system) without any dehydration. Further investigations clarified that the dehydration system was actually shut down a few years ago based on a recommendation made by a (allegedly) visiting process engineer. He had referred to the available lab reports, which indicated that the gas received at the terminal was dry. Therefore, he concluded that significant cost savings can be made by shutting down the dehydration system! When the Team reviewed such old lab reports it became evident that somehow the lab (which was analyzing gas samples from the gas terminal on a regular basis) would not report the water content of the gas in its reports. Hence, some personnel had the impression that the gas that was flowing through their terminal was dry!

Culprit: Total lack of competency among those responsible for operating the gas terminal in tandem with the incompetence of the visiting process engineer who concluded that the sour gas was dry (i.e., lack of competency at different levels and departments).

Outcome: Huge annual repair and maintenance costs caused by recurring leaks downstream of the dehydration system shortly after it was shut down.

Corrosion Management Recommendations: The main recommendation was to fully rehabilitate and reinstate the gas dehydration system to ensure it would dehydrate the sour wet gas per the required specifications. Furthermore, a set of corrosion control matrices and corrosion key performance indicators were produced for the dehydration unit to ensure it would function more efficiently.

Case Study 12: A Subsea Gas Pipeline

Overview: A subsea carbon steel pipeline has been used for a long time for transferring sweet wet gas from a number of subsea wells to an onshore terminal. Because both the CO_2 and the water content of the gas have been low since commissioning, the failure risk due to corrosion has been deemed negligible. Thus, because of this low failure risk, no monitoring or intelligent pigging has been carried out for the pipeline for many years. However, after more than 20 years of impeccable operation it was decided to carry out an intelligent pigging of the pipeline to further confirm its pristine integrity conditions. Surprisingly, the results show that there is hardly any wall thickness left at certain points of the pipeline and it has to be shut down.

Observation: The CM Team went through the available data and confirmed that the internal corrosion threat to the pipeline was very low from day one because of the low water cut and the negligible carbon dioxide content. However, further investigations revealed that the methanol that was being injected into the pipeline inlet (to prevent hydrate formation) was coming from a tank that was not purged (with nitrogen) for a number of years. Thus, the oxygen ingress into the methanol injection system carried over to the pipeline and caused severe localized corrosion at lower points where water could drop out or accumulate in the pipeline.

Culprit: Purging the methanol tank was discontinued at some point, either to save money or simply because of negligence or lack of competency (i.e., more probably total lack of competency on the operator side).

Outcome: Total shutdown of the pipeline, causing the operator considerable financial loss and also leaving them with a tarnished reputation.

Corrosion Management Recommendations: The main recommendation was to review all other pipelines' chemical treatment history to find out whether there were any similar cases. It was also recommended to add a new item to each pipeline CCM Document under "nitrogen purging" to make this activity more visible.

Case Study 13: A Natural Gas Distribution Company (Downstream Sector)

Overview: A natural gas distribution company (downstream sector) receives dried natural gas from a multitude of gas refineries (upstream sector) and then distributes the received processed gas among its numerous household and industrial clients through a buried pipeline distribution network extending more than 12,000 km.

Because the gas is dried, no internal deterioration or corrosion is expected and externally the buried pipeline network is protected by the combination of cathodic protection and coating applications. Shortly after commissioning new sections, the pipes start leaking. Simultaneously, some older sections develop leaks; requiring the CM Team to carry out the relevant failure investigations.

Observation: The reviews carried out as part of the ongoing failure investigations revealed that the new sections were handed over to the Operations Team at commissioning without removing the hydrotest water from those sections. The leaks associated with older sections were due to water ingress into the buried pipeline network at excavations where the water in the excavated ditch was not removed per the existing procedures before opening up the system. This resulted in water ingress into the pipeline network at several excavated locations.

Culprit: Not adhering to the existing hydrotest and repair procedures that demanded hydrotest water removal shortly after completing the hydrotest activities and using a pump to remove collected water in excavated ditches prior to opening up the system, respectively.

Outcome: Frequent gas leaks due to internal corrosion resulting in significant repair and replacement costs along with interruption in providing gas to various household and industrial clients.

Corrosion Management Recommendations: The main recommendation was to ensure that both the hydrotest procedure and the water removal procedure for excavated ditches are strictly adhered to in future cases. Furthermore, the operator was to provide extra training for relevant contractor and company personnel to ensure that such issues would not happen again in future.

Case Study 14: An Overland Oil Pipeline

Overview: A newly built and commissioned overland wet oil pipeline suffers from multiple failures during hydrotesting. Upon rectifying such failures and entering service, numerous corrosion leaks and failures are identified along the pipeline length.

Observation: The CM Team reviewed the pipeline leak data and within the first few years of commissioning they amounted to hundreds of corrosion leaks and failures. Further investigations revealed that the pipeline's design corrosion allowance was used up prior to hydrotesting at numerous locations.

Culprit: The pipeline sections arrived by ship and the sections that were stored in lower ship compartments were totally or partially immersed in seawater, which was present in those compartments. This led to severe corrosion of the brand new sections. Not following existing storage and transportation procedures was deemed the main culprit.

Outcome: Frequent oil pipeline shutdown associated with increased repair and maintenance costs, along with regular shutdown of the downstream facility that received its feed oil through the pipeline.

Corrosion Management Recommendations: The CM Team's main recommendation was to create proper transportation and storage guidelines and procedures requiring improved pipeline sections transport and storage routines. Also, communication had to be improved between the purchasing company and those responsible for the storage and transportation of the purchased goods, in addition to improved inspection and supervision during transportation.

Case Study 15: An Onshore Seawater Injection System

Overview: An onshore seawater injection system takes seawater from the sea and after treatment (oxygen removal, chlorination, and biociding) sends the treated seawater along a subsea pipeline to an offshore platform for its injection into a multitude of subsea injection wells. The whole system is manufactured from carbon steel and the main internal protection against corrosion is proper seawater treatment.

Observation: A CM consultant was brought in to carry out an integrity review of the seawater injection system concerned. He noticed odd chemical treatment practices such as chlorination and batch biociding *upstream* of the deaeration, along with batch biociding *downstream* of the deaeration. He also noticed continuous scale inhibitor and corrosion inhibitor injections upstream and downstream of the filters, respectively. The chemical treatment had been in place for around 20 years. However, considering the available integrity information, there was no justification for batch biociding upstream of the filters as well as the continuous injection of scale and corrosion inhibitors..

Culprit: Total lack of competency among operator personnel for having continued unnecessary chemical treatments for 20 years. However, although they could not justify such erroneous chemical treatment, they would have continued to do so merely because of the fact that their predecessors also carried out similar chemical treatments!

Outcome: Bearing huge unnecessary chemical treatment costs for 20 years (and still continuing with it!) where they could have spent the money more cleverly or could have simply made huge savings by refraining from injecting unnecessary chemicals.

Corrosion Management Recommendations: The main recommendation was to provide the necessary training courses for the pertinent operations personnel so they could make better and more efficient decisions regarding the incumbent chemical treatment system associated with the above seawater injection system.

Case Study 16: An Oil Storage and Export Terminal

Overview: An oil storage and export terminal has been in operation for more than 22 years. Its piping and various equipment are mostly made of carbon steel, and the internal corrosion threat is mitigated through corrosion inhibitor injection at various locations. Because of an effective integrity management program, the number of corrosion leaks and failures has been very few and almost negligible over the past two decades.

Observation: Corrosion leaks and failures started to appear at an increasing rate virtually out of nowhere. UT wall thickness inspections identified numerous locations with severe localized corrosion. Because the main operations and process parameters had remained the same, a failure investigation team was formed to identify the culprit behind such abrupt appearance of corrosion leaks at various locations.

Culprit: The ongoing investigations revealed that the recent change-out of the corrosion inhibitor type was the main cause. A new chemical had been purchased to replace an old corrosion inhibitor and the purchased chemical was actually an acidic fluid that should never come into contact with carbon steel.

Outcome: Huge repair and replacement costs along with environmental contamination by the leaked oil and frequent asset shutdowns leading to reduced export rates.

Corrosion Management Recommendations: The CM Team reviewed and revised the incumbent chemical purchasing and testing procedures. It was also recommended that all nominated chemicals should be tested in an independent third party lab to ensure their suitability for the facilities concerned.

Corrosion Management Exercises

Introduction: The following exercises are intended to present the reader with an opportunity to test their corrosion management knowledge and use that knowledge in a more practical and effective manner. The associated answers are provided at the end of the exercise questions.

Exercises:

1. What is the definition of an "Integrity Management Measure"?

2. What is the definition of the "Corrosion Management" process according to MCMM?

3. According to Figure 3.2 how many integrity management measures are there?

4. What are the main parameters affecting the "Inspection Requirements"?

5. What are the main parameters affecting the "Corrosion Rate Monitoring Requirements"?

6. An asset corrosion engineer has been tasked with producing a fluid sampling system for an offshore platform producing wet sweet oil and gas. What parameters should be taken into consideration when producing the required sampling system?

7. An asset corrosion engineer has been tasked with devising a corrosion inhibitor injection system. What parameters should be taken into account when doing so?

8. What are the various data types that can be used as input to the failure risk assessment process?

9. An asset suffers from frequent corrosion failures. To take advantage of the failure data it has been decided to produce a Leak Register Database. Determine at least 10 leak parameters to be included in this register.

10. Create two lists of variables that can be used to determine probability and Consequence of failure for a buried gas distribution system serving a city.

Answers:

1. Anything that contributes to the overall integrity management (IM) system or influences its functioning and outcome. Integrity management measures are either CE-based or non-CE-based.

2. The process of reviewing the required IM measures, regular monitoring of their performance, and the assessment of their effectiveness post-commissioning.

3. According to Figure 3.2 there is a total of 18 integrity management measures: 3 CE-based and 15 non-CE-based ones.

4. They include
   - Inspection technique (e.g., UT wall thickness inspection, radiography, and visual)
   - Inspection type (e.g., time-based, condition-based, or risk-based)
   - Proper application or implementation of the chosen inspection technique
   - Inspection frequency
   - Susceptible features
   - Inspection area or extent
   - Data reliability
   - Data processing
   - Dissemination of findings
   - Competency of those involved in the inspection process

5. They include
   - Corrosion rate monitoring system type (e.g., probes or coupons)
   - Location
   - Orientation
   - Service requirements
   - Data reliability
   - Data processing
   - Competency of those involved

6. The parameters that should be taken into consideration by the asset corrosion engineer
   - Sampling location
   - Sampling frequency
   - Sampling procedure
   - Sampling timing with respect to chemical treatment (e.g., fluid sampling for determining bacterial numbers should be carried out both before and after biociding)
   - Data reliability
   - Data processing
   - Independence of the lab where the samples are analyzed
   - Competency of those involved

7. The parameters that should be taken into consideration by the asset corrosion engineer
    - Whether corrosion inhibitor injection is at all required (i.e., in case of low water levels and, depending on the prevailing flow conditions, water wetting of the internal surfaces might not occur; thus no corrosion inhibitor injection is required)
    - Injection location
    - Injection frequency (i.e., continuous or batch)
    - Application method (e.g., whether injection quills are used at the injection points)
    - Chemical type
    - Injection rate or concentration of the injected chemical
    - Weld protection (i.e., some chemicals do not provide the same level of protection to weld and heat affected zone areas)
    - Presence of sand (thus preventing the chemical to be adsorbed onto the surface where the surface is covered by sand)
    - Chemical compatibility with other injected chemicals
    - Chemical compatibility with metals/alloys concerned
    - Process conditions (e.g., too high temperatures can disintegrate or neutralize some corrosion inhibitors)

8. Such data types can include
    - Design
    - Operations
    - Process
    - Chemical treatment
    - Corrosion rate monitoring
    - Fluid sampling
    - Cathodic protection
    - Coating
    - Repairs (because of corrosion)
    - Anomalies
    - Leaks

9. "Leak Register" parameters can include
    - Date (i.e., leak date)
    - Type of item leaked (e.g., pipeline, piping, weld, and bend)
    - Leak orientation
    - Leak location with respect to most immediate upstream equipment (e.g., pump, heater, separator, injection location, and tank)
    - Vicinity to nearest weld
    - Material or alloy type
    - Mode of failure (e.g. pitting or rupture)
    - Internal or external corrosive environment causing the leak
    - Last inspection date
    - Design life versus actual life
    - Deterioration rate

- Corrosion mitigation method at the leak area (e.g., design, materials selection, or chemical injection)

10. Variables that influence the probability of failure can include
    - Corrosion rate
    - Number of leaks
    - Coating condition
    - Cathodic protection system performance
    - Number of repairs and anomalies (due to corrosion)
    - Corrosivity of internal and external environments
    - Water content
    - And variables that influence the consequence of failure can include
        - Pipeline diameter
        - Pipeline pressure
        - Type of fluid inside
        - Vicinity to population centers
        - Population size
        - Extent of shutdown
        - Repair costs

Abbreviations and Acronyms

AIM	Asset Integrity Management
AIMS	Asset Integrity Management System
API RP	American Petroleum Institute Recommended Practice
ASME	American Society of Mechanical Engineers
CCM	Corrosion Control Matrices
CE	Corrosion Engineering
CM	Corrosion Management
CML	Commercial Microbiology Limited
CMS	Corrosion Management Strategy
CNC	Criteria of Non-Conformance
CP	Cathodic Protection
CRA	Corrosion Risk Assessment
CRA	Corrosion Resistant Alloy
CUI	Corrosion Under Insulation
DNV	Det Norske Veritas
EMEA	Europe, Middle East, and Africa
EXT	External
FRA	Failure Risk Assessment
GDP	Gross Domestic Product
GVI	General Visual Inspection

HAZ	Hazard
HP	High Pressure
HSE	Health and Safety Executive
IM	Integrity Management
INT	Internal
IRP	Integrity Review Process
KPIs	Key Performance Indicators
LP	Low Pressure
LR	Lloyd's Register
MAWT	Minimum Allowable Wall Thickness
MCMM	Morshed Corrosion Management Model
MIC	Microbially Induced Corrosion
MP	Materials Performance
MSD	Materials Selection Diagram
NACE	National Association of Corrosion Engineers
OPER	Operability
PFD	Process Flow Diagram
P&IDs	Piping and Instrumentation Diagrams
ppb	Parts per billion
ppm	Parts per million
PSN	Production Services Network
PSV	Pressure Safety Valve
RBI	Risk-Based Inspection
SRB	Sulfate-Reducing Bacteria
UT	Ultrasonic Testing

Index

D

E

www.ingramcontent.com/pod-product-compliance
Ingram Content Group UK Ltd.
Pitfield, Milton Keynes, MK11 3LW, UK
UKHW020159290926
15471UKWH00005B/961

9 781575 904078

NACE INTERNATIONAL

ISBN 978-1-57590-407-8
90000
9 781575 904078